BIOCHEMISTRY RESEARCH TRENDS

DEHYDROGENASES

BIOCHEMISTRY, FUNCTIONS AND ROLE IN DISEASE

BIOCHEMISTRY RESEARCH TRENDS

Additional books in this series can be found on Nova's website under the Series tab.

Additional e-books in this series can be found on Nova's website under the eBook tab.

BIOCHEMISTRY RESEARCH TRENDS

DEHYDROGENASES

BIOCHEMISTRY, FUNCTIONS AND ROLE IN DISEASE

MARGARET COLLINS
EDITOR

nova publishers
New York

Library of Congress Cataloging-in-Publication Data

ISBN: 978-1-53610-550-6

Library of Congress Control Number: 2016958943

Published by Nova Science Publishers, Inc. † New York

CONTENTS

PREFACE

Dehydrogenases are enzymes that play an important role in oxidation reactions. Dehydrogenases are involved in many metabolic pathways which maintain normal function and homeostasis of cells in the body. This book discusses the biochemistry of dehydrogenases, as well as their functions and roles in diseases.

Chapter 1 – Lactate dehydrogenase (EC 1.1.1.27, LDH) is an enzyme ubiquitary distributed in cells of vertebrates, plants, and bacteria. It takes place in metabolism of carbohydrates catalyzing interconversion of lactate and pyruvate with nicotinamide adenine dinucleotide as coenzyme both in cytoplasm as well as in mitochondria. Structurally, it is a tetramer of four units (M_r 35,000 each), which in animals exists in five electrophoretically distinguishable forms known as isoenzymes. It was shown that bird isoenzymes differ from their mammalian counterparts in their electrophoretical properties, especially in their pI values. Literature sources and reports dealing with the analysis of LDH from the aspect of the physiology of animals and pathophysiology of metabolic and systemic disorders also demonstrate that the activity of LDH and its serum and tissue patterns show a great variations between the animal species. Therefore, it is necessary to understand the LDH composition and isoenzyme distribution in respective animals as these differences do not allow to use catalytic activities of LDH isoenzymes from one species to another that is important in particular in veterinary medicine. Even though LDH and its isoenzymes examination are not routinely used in veterinary laboratory practice many reports suggest their usefulness as a valuable adjunct indicator in diagnostics of animal diseases.

Chapter 2 – In kidney the role of urea is to modulate urine concentration, however, urea is a protein denaturant with a strong effect on several renal enzymes. The enzyme betaine aldehyde dehydrogenase catalyzes the synthesis of glycine betaine and is located in cortex and medulla cells of kidney, and is therefore, exposed to high urea concentrations. Betaine aldehyde dehydrogenase follows an ordered sequential kinetic mechanism with NAD^+ as the first substrate to bind to the enzyme. The effect of urea in the BADH kinetic parameters was analyzed under two ionic strength conditions. Enzyme activity was assayed at 0-2 M concentrations of urea and pH 7.0. The results showed that urea was a competitive inhibitor with respect to NAD^+ and non-competitive regarding the betaine aldehyde substrate. The competitive and non-competitive inhibition constants were calculated. Initial velocity patterns in presence of 150 mM NaCl or KCl showed that both salts increased the affinity of BADH for NAD^+. These results suggest that enzyme inhibition by urea is reversed by intracellular potassium concentrations, ensuring the synthesis of glycine betaine; an important osmolyte and osmoprotector in kidney. During the mechanism of urine concentration, the activity of betaine aldehyde dehydrogenase is maintained despite the high urea concentrations because of the hyperosmolarity required to produce concentrated urine.

Chapter 3 – Dehydrogenases are enzymes that play an important role in oxidation reactions. They oxidize a specific substrate in which one or more hydride ions are transferred to an electron acceptor molecule. Dehydrogenases are involved in many metabolic pathways which maintain normal function and homeostasis of cells in the body. One example is glucose metabolism, which is crucial for providing energy (ATP) as well as other molecules, such as reduced form of nicotinamide adenine dinucleotide phosphate (NADPH) and ribose sugar. In this chapter, the authors review the 10 dehydrogenase enzymes found in the main pathways of glucose metabolism: glycolysis, lactic acid fermentation, alcoholic fermentation, the citric acid cycle and pentose phosphate pathway. The authors also describe the role of these dehydrogenase enzymes in diseases. Glucose metabolism is essential to provide metabolic energy to many cells and tissues. Therefore the impairment of function, activity or an alteration in the level of expression of these 10 dehydrogenase enzymes has an impact on the function of many cells and tissues. Convincing evidence has indicated dehydrogenase enzymes are associated with several diseases, including neurological disorders, many types of cancer and blood diseases. In this chapter, they also discuss how dehydrogenases contribute to the etiology of diseases and their clinical manifestations. Additionally, the authors explore the applications and advantages of using dehydrogenase enzymes in current medical diagnosis and treatment.

In: Dehydrogenases
Editor: Margaret Collins

ISBN: 978-1-53610-550-6

Chapter 1

LACTATE DEHYDROGENASE OF BIRD AND MAMMALIAN ORIGIN

Dagmar Heinová[1], Zuzana Kostecká[1,*] and Oskar Nagy[2]
[1]Department of Chemistry, Biochemistry and Biophysics
[2]Clinic of Ruminants,
University of Veterinary Medicine and Pharmacy, Košice, Slovakia

ABSTRACT

Lactate dehydrogenase (EC 1.1.1.27, LDH) is an enzyme ubiquitary distributed in cells of vertebrates, plants, and bacteria. It takes place in metabolism of carbohydrates catalyzing interconversion of lactate and pyruvate with nicotinamide adenine dinucleotide as coenzyme both in cytoplasm as well as in mitochondria. Structurally, it is a tetramer of four units (M_r 35,000 each), which in animals exists in five electrophoretically distinguishable forms known as isoenzymes. It was shown that bird isoenzymes differ from their mammalian counterparts in their electrophoretical properties, especially in their pI values. Literature sources and reports dealing with the analysis of LDH from the aspect of the physiology of animals and pathophysiology of metabolic and systemic disorders also demonstrate that the activity of LDH and its serum and tissue patterns show a great variations between the animal species. Therefore, it is necessary to understand the LDH composition and isoenzyme distribution in respective animals as these differences do not allow to use catalytic activities of LDH isoenzymes from one species

* Corresponding Author E-mail: zuzana.kostecka@uvlf.sk.

to another that is important in particular in veterinary medicine. Even though LDH and its isoenzymes examination are not routinely used in veterinary laboratory practice many reports suggest their usefulness as a valuable adjunct indicator in diagnostics of animal diseases.

Keywords: lactate dehydrogenase, isoenzymes, birds, mammals

INTRODUCTION

The value of serum enzyme assays as indicators of the site of tissue damage is limited by the fact that only a few enzymes are tissue specific. In the second half of the last century an attention of biochemists turned to the possible application of isoenzymes in the assessment of organ disorders as some of them distinguish by specific isoenzyme profile. Isoenzymes, as it is known, are genetically determined structurally different forms of enzymes (Li et al., 1990). One of them is lactate dehydrogenase (L-lactate:NAD oxidoreductase, EC 1.1.1.27, LDH), a tetrameric enzyme, which exists in three basic homotetrameric forms: H_4 (LDH_1), M_4 (LDH_5), and C_4 (LDH-X) in vertebrates (Cahn et al., 1962; Wilson et al., 1963; Roussel & Stallcup, 1967; Holmes, 1972; Wu et al., 1987; Li et al., 1990; Quattro et al., 1993; Goldberg et al., 2010). Three structurally different polypeptide chains of homotetrameric LDH molecules are encoded by three different genes (Zinkham & Isensee, 1972; Wu et al., 1987; Selehi-Ashtiani & Goldberg, 1993; Tsoi & Li, 1994; Goldberg et al., 2010).

Lactate dehydrogenase is widely distributed in cells of various living systems where it is involved in carbohydrate metabolism catalyzing interconversion of lactate and pyruvate with NAD^+/NADH coenzyme system. In the cell, it is localized in cytoplasm (cLDH) where it produces lactate from pyruvate – the end product of glycolysis. Except for cytosol, lactate dehydrogenase is also localized inside the outer mitochondrial membrane (mLDH) (Elustondo et al., 2013) where it likely oxidizes lactate as predominant monocarboxylate (Brooks et al., 1999). This is in accordance with the hypothesis of intracellular lactate shuttle between the cytosol and mitochondria according to which lactate generated in cytosol is shuttled into mitochondrion of the same cell via mitochondrial monocarboxylate transporter (Brooks, 1999; Elustondo et al., 2013). In the mitochondrion, lactate is oxidized by mLDH into pyruvate for tricarboxylic acid cycle (Schurr, 2014).

Lactate dehydrogenase H_4 and M_4 homotetramers, also called somatic LDH (Bonny et al., 1998), are present in tissues of all studied vertebrates while the tissue distribution of the LDH-C_4 isoenzyme varies from organism to organism. In lower teleost fishes, LDH-C has a generalized tissue distribution, but in advanced teleost fishes it is found either in the liver (e.g., in cod) or in the eye (e.g., in salmon). In mammals and columbid birds, the LDH-C_4 isoenzyme is expressed only in mature testes (Tsuji et al., 1994; Tsoi & Li, 1994; Mannen et al., 1997). Somatic isoenzymes of lactate dehydrogenase exist in five electrophoretically distinguishable forms. According to their different mobility to the anode they are denoted LDH_1 (H_4), LDH_2 (H_3M), LDH_3 (H_2M_2), LDH_4 (HM_3) and LDH_5 (M_4) with LDH_1 having the highest and LDH_5 the lowest migration rate to the anode. Three hybrid forms (LDH_2, LDH_3, and LDH_4) are relatively equally spread between LDH_1 and LDH_5 on the electrophoreogram. Protomers of the somatic LDH isoenzymes are, in general, designated H (heart) or B, and M (muscle) or A (Tsuji et al., 1994; Mannen et al., 1997; Singh et al., 2012) according to the domination of homotetrameric molecules in the cells of respective organs of adult vertebrates. Hybrid forms of LDH are present in various levels in the individual organs of an animal/human.

Lactate dehydrogenase isoenzymes catalyze the same biochemical reaction (Plummer & Wilkinson, 1962; Battelino & Blanco, 1970), all have the same molecular weight of 134,000 (Appella & Markert, 1961), and all contain four polypeptide chains each of molecular weight of 33,500 (Cahn et al., 1962; Battelino & Blanco, 1970). Two different kinds of polypeptide chains differ in their amino acid sequences and composition with 333 amino acid residues in each chain (Fondy et al., 1964; Fondy & Kaplan, 1965; Wu et al., 1987; Tsoi & Li, 1994; Tsuji et al., 1994). Although catalyzing the same overall reaction, they differ in their kinetic properties (K_m and k_{cat}) (Table 1) (Plagemann et al., 1960b; Dawson et al., 1964; Fondy & Kaplan, 1965; Battelino & Blanco, 1970; Kopperschlager & Kirchberger, 1996; Holland et al., 1997), inhibition by substrate (pyruvate) (Dawson et al., 1964; Fondy & Kaplan, 1965), and immunological response (Plagemann et al., 1960a; Dawson et al., 1964).

Different structure of lactate dehydrogenase isoenzyme molecules predetermines their different net charge, and consequently, different migration rate in the electric field which is used in separating of these enzymes in research as well as in diagnostic practice. The LDH isoenzyme system has become one of the most extensively studied models used to investigate the origin and evolution of isoenzymes and regulation of multigene families (Holmes 1972; Markert et al., 1975; Li et al., 1983; Quattro et al., 1993; Tsoi & Li, 1994; Tsuji et al., 1994; Mannen et al., 1997; Li et al., 2002).

Table 1. Kinetic constants K_m, k_{cat} and C (k_{cat}/K_m) determined for the conversion of pyruvate into lactate by the catalytic action of H_4 and M_4 lactate dehydrogenase homotetramers

Origin	K_m (mol·l^{-1})		k_{cat} (s^{-1})		C (s^{-1}·l·mol^{-1})	
	H_4	M_4	H_4	M_4	H_4	M_4
Chicken	$8.9 \cdot 10^{-5}$	$3.2 \cdot 10^{-3}$	45,000	93,400	$5 \cdot 10^{8}$	$2.9 \cdot 10^{7}$
Rabbit	$6.7 \cdot 10^{-5}$	$3.5 \cdot 10^{-4}$	41,000	?	$6 \cdot 10^{8}$	?
Cattle	$7.1 \cdot 10^{-5}$	$1.0 \cdot 10^{-4}$	49,400	80,200	$7 \cdot 10^{8}$	$8.0 \cdot 10^{8}$

Electrophoretic Separation of Bird and Mammalian Lactate Dehydrogenases

Determination of total LDH activity and its isoenzyme pattern in serum of mammals had soon become one of the biochemical indicators in the assessment of organ disorders. The application of bird lactate dehydrogenases in research as well as in diagnostic practice remained limited to the determination of total serum activity as the use of native electrophoresis did not allow isoenzyme separation. Using isoelectric focusing technique in a gradient of pH 3 to 9 (Heinová et al., 1996; Heinová et al., 2016) enabled good and clear resolution of all five bird LDH fractions and revealed the reason of their immobility in the electric field with a buffer system of the pH values 8.6 – 8.8, i.e., under condition commonly used for the separation of mammalian lactate dehydrogenases. At pH 8.6, the electrophoretic migration of LDH isoenzymes depends on the migration of the two pure types, i.e., the H_4 and M_4 forms (Cahn et al., 1962). The more widely the two homotetramers differ in charge (the case of mammalian LDHs), the more separable are hybrids by electrophoresis. In the case of bird LDHs, the two pure types migrate relatively close together toward the negative pole at pH 8.6, and observation of the hybrid forms under these conditions is very difficult (Cahn et al., 1962). Localization of bird LDH isoenzymes on the electrophoreogram after separation in the pH gradient of 3 to 9 revealed that the H_4 and M_4 molecules migrated close together to the cathode and were localized in the pH region of 6.2 to 8.1 (chicken LDH isoenzymes), 5.4 to 7.7 (pheasant LDHs), and 5.3 to 8.3 (turkey lactate dehydrogenase isoenzymes) (Heinová et al., 1996). Mammalian molecules of LDH were more acidic and widespread with the pI values being in the range of 4.5 to 9.0 (Table 2).

Table 2. Isoelectric point values of lactate dehydrogenase isoenzymes of some birds and mammals

Origin	Isoelectric point pI				
	LDH_1	LDH_2	LDH_3	LDH_4	LDH_5
Chicken	6.2	7.0	7.4	7.7	8.1
Pheasant	5.4	6.3	6.8	7.3	7.7
Turkey	5.3	6.2	7.0	7.6	8.3
Horse	4.8	5.5	6.8	?	9.0
Cattle	4.5	4.9	5.6	6.7	?
Rabbit	4.5	5.1	5.8	6.7	9.0

From the literature it is known that bird and mammalian lactate dehydrogenases differ in the amino acid composition of their H and M subunits (Fondy et al., 1964; Fondy & Kaplan, 1965; Tsoi & Li, 1994; Tsuji et al., 1994; Mannen et al., 1997). Comparing avian and mammalian enzymes, the avian H, and especially M molecule contain more basic amino acid residues, in particular histidine (Fondy & Kaplan, 1965). This can explain the different electrophoretic mobilities of bird and mammalian LDH isoenzymes in a pH gradient of 3 to 9. An apparent shift of the isoelectric points of bird isoenzymes toward basic values can also explain the unwillingness of bird LDHs to separate at pH 8.6 using native electrophoresis and producing only one enzymatic zone (Heinová et al., 1996; Heinová et al., 2016).

From studies on the evolution and phylogenetic relationship of species it is known that heart and skeletal muscle lactate dehydrogenases of birds and mammals have similar catalytic properties (Plagemann et al., 1960b; Dawson et al., 1964; Fondy & Kaplan, 1965) (Table 1). Therefore, it is probable that different electrophoretic properties of LDH isoenzymes reflect the different phylogenesis of bird and mammalian LDH molecules that is validated by molecular studies of species evolution (Hedges, 1994; Tsuji et al., 1994; Mannen et al., 1997; Li et al., 2002).

Organ and Serum Pattern of Lactate Dehydrogenases in Different Species

Cells of tissues are direct source of lactate dehydrogenase isoenzymes that are naturally distributed in blood plasma/serum of animals and humans producing characteristic profile (Plagemann et al., 1960a, 1960b; Boyd, 1962;

Thorton & Lohni, 1979; Beatty & Doxey, 1983; Milne & Doxey, 1987; Maekawa, 1988; Preus et al., 1989; Yasuda et al., 1990; Heinová et al., 1999; Nagy et al., 2011; Heinová et al., 2016) (Figure 1). This profile depends on intracellular isoenzyme concentration in all tissues that contribute to the common pool of lactate dehydrogenases in plasma/serum as a consequence of natural cell degradation. When the content of cells is released from tissue to plasma, as on cell injury, the LDH isoenzyme pattern of the serum changes in favour of the profile of the affected organ (Maekawa, 1988) that can be used in the diagnostic practice.

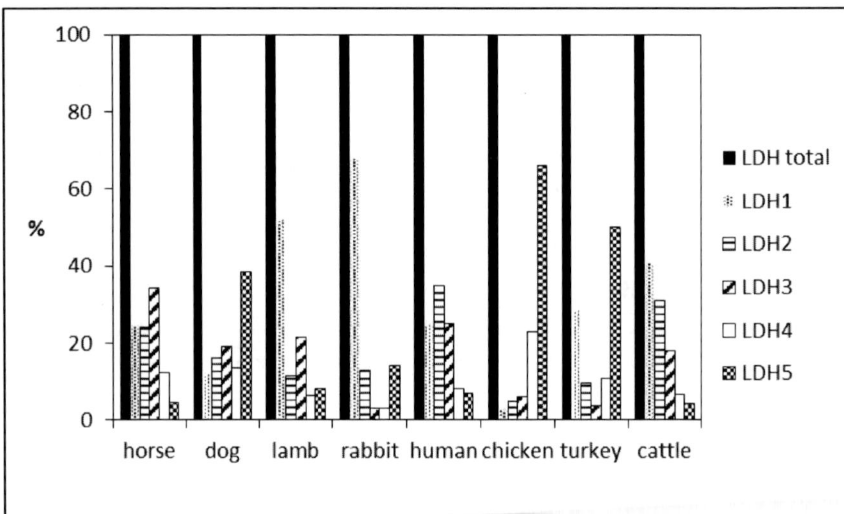

Figure 1. Relative distribution (%) of lactate dehydrogenase isoenzymes in sera of some birds and mammals.

Generally, heart, skeletal muscle, and liver are LDH rich organs that differ from each other in their isoenzyme pattern (Plagemann et al., 1960a; Thorton & Lohni, 1979; Beatty & Doxey, 1983; Milne & Doxey, 1987; Heinová et al., 1999; Singh et al., 2012; Heinová et al., 2016). In cardiac muscle, anodic LDH_1 and LDH_2 dominate over cathodic forms of LDH_5 and LDH_4 (Figure 2) while in skeletal muscle of most vertebrates the pattern is opposite with a prevalence of cathodic isoenzymes (Figure 3). In other organs, hybrid forms are also present with a more uniform distribution and a relatively lower catalytic intensity. Therefore, disorder to the integrity of heart cells, e.g., during heart attack is accompanied with increased level of anodic fractions, especially LDH_1 in the serum of an affected individual (Maekawa, 1988;

Aguilera-Tejero et al., 1989; Huijgen et al., 1997). In the case of the injury to the skeletal muscle cells the increased activity of LDH_5 in serum is observed (Aguilera-Tejero et al., 1989; Huijgen et al., 1997; Bogin 1994; Cardinett III, 1997).

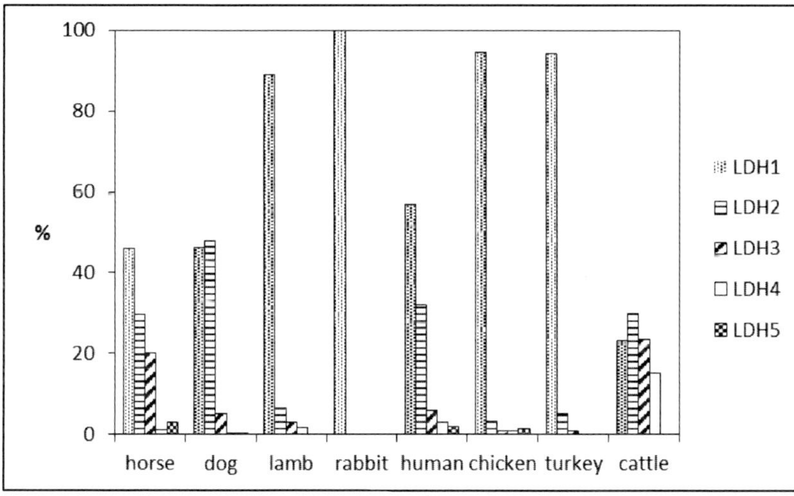

Figure 2. Relative distribution (%) of lactate dehydrogenase isoenzymes in heart muscle of some birds and mammals.

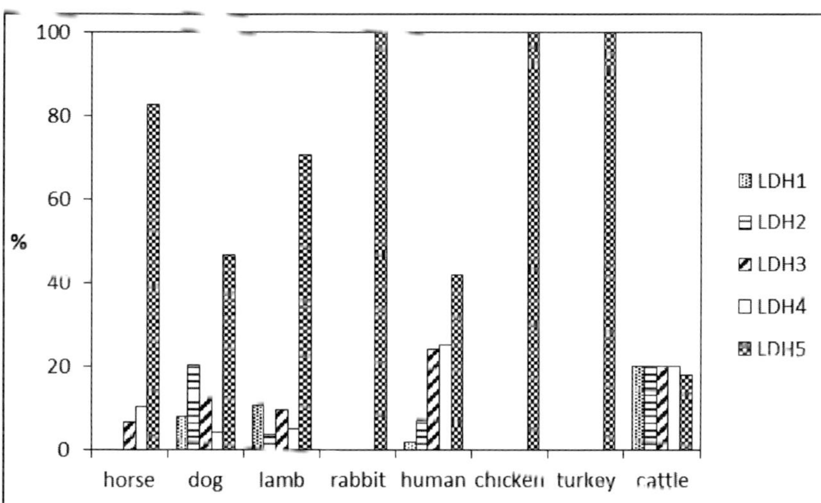

Figure 3. Relative distribution (%) of lactate dehydrogenase isoenzymes in skeletal muscle of some birds and mammals.

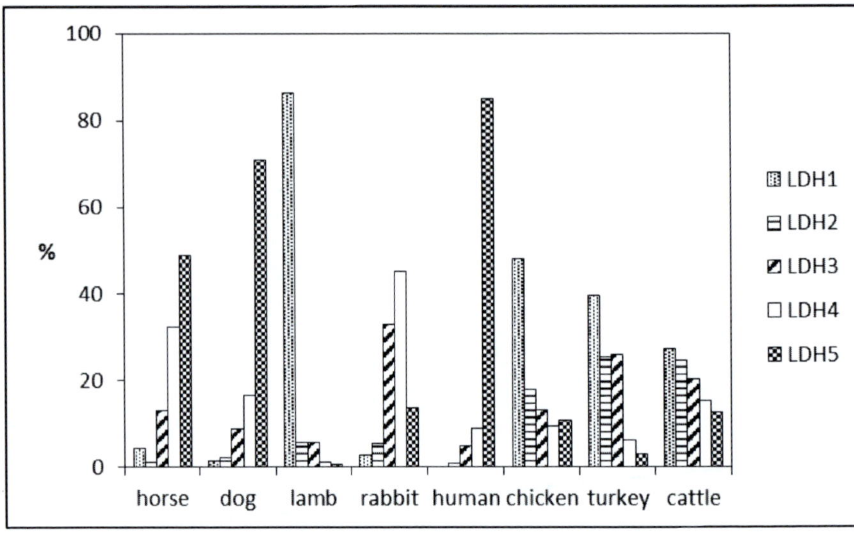

Figure 4. Relative distribution (%) of lactate dehydrogenase isoenzymes in liver of some birds and mammals.

The interpretation of a changed pattern of serum LDH isoenzymes in relationship to the assessment of other organ disorders, especially liver, remains, however, questionable. Generally, an idea is accepted that serum LDH_5 is a so called liver enzyme (liver LDH), and some authors even designated it LDH-LL. This can lead to an incorrect conclusion that LDH_5 can serve as a general indicator of damage to the membrane of liver cells which is not consistent with the diversity of LDH_5 level in liver of different animal species (Figure 4). As seen, three groups of LDH isoenzymes in liver can be observed: a) a group with the domination of cathodic LDH_5 (humans, dog, horse); b) that with anodic LDH_1 domination (lamb, chicken, turkey), and c) that with the prevalence of hybrid forms (rabbit, cattle). Moreover, skeletal muscle cells quantitatively dominate over the liver cells in respect to the level of LDH_5 as documented in the Table 3. (The comparison of the LDH_5 levels in the Table 3 was obtained as a ratio of skeletal muscle LDH_5 specific activity (U/g) to the specific LDH_5 activity calculated for liver LDH_5.). Skeletal muscle is not only the richest tissue on LDH_5 isoenzyme but in the same time it represents the major mass component of the body that can influence the serum LDH profile as well. Therefore, for the interpretation of LDH isoenzyme pattern in the serum diagnostics of the liver and muscle disorders it is more convenient to use the principle of isoenzyme different electrophoretic mobility to the anode, i.e., LDH_1, LDH_2, LDH_3, LDH_4 and LDH_5.

Table 3. The ratio of LDH_5 isoenzyme concentration in skeletal muscle to the concentration in liver calculated for some animals

	Horse	Dog	Lamb	Rabbit	Chicken	Turkey	Cattle
Skeletal muscle	T	T	T	T	T	T	T
	11.7x	4.3x	17.5x	6.2x	76.8x	4,900x	1.2x
Liver	⊥	⊥	⊥	⊥	⊥	⊥	⊥

Other organs with all five isoenzyme forms contribute only in the relatively low concentrations to the total pool of serum LDH (Plagemann et al., 1960a; Thorton & Lohni, 1979; Beatty & Doxey, 1983; Milne & Doxey, 1987; Heinová et al., 1999; Singh et al., 2012; Heinová et al., 2016). These lactate dehydrogenase isoforms probably come from the cells of all animal organs and their tissue origin in the serum thus cannot be identified with certainty.

The activity of LDH and its serum and tissue patterns and composition show a great variations between the species (Figures 1 – 4). These differences do not allow to use catalytic activities of LDH isoenzymes from one species to another. Instead, the pattern of serum LDH isoenzymes should be interpreted in respect to its species origin that is important in particular in veterinary medicine.

CLINICAL APPLICATIONS OF LACTATE DEHYDROGENASE

Serum enzymes, including lactate dehydrogenase, play an important role in diagnosis of organ disorders or damage in animals and humans. Increase in LDH has been reported in patients with various diseases with the more or less characteristic changes in the LDH isoenzymes pattern. In addition to commonly used blood serum or plasma, LDH activities were also analysed in other body fluids of animals, e.g., synovial fluid (Schmökel et al., 2001), cerebrospinal fluid (Nečas & Sedláková, 1999), milk (Bogin et al., 1977; Babaei et al., 2007), and bronchoalveolar lavage fluid (Davies et al., 1986; Weiss et al., 1991; Reinhold et al., 1992).

The widespread occurance of LDH isoenzymes in the animal organs makes this enzyme not specific with the total blood serum activity increased in e.g., muscular dystrophy, liver necrosis, cancer and various disorders of other organ systems. Determination of other enzymes activities thus should be taken into account when diagnosing disease and supporting therapeutic decision or predicting prognosis (Tzouvelekis et al., 2005). Skeletal and cardiac muscle cells contain high concentrations of lactate dehydrogenase. The dominating isoenzyme in the heart is LDH_1 and LDH_2; in skeletal muscle the dominating isoenzyme is LDH_5. Blood levels of these LDH isoenzymes are therefore monitored for the evaluation of physical fitness and degree of training (Anderson, 1976; Bogin, 1994; Chaplin et al., 1997; Heinová & Vaško, 2007). Since hepatocytes also contain large concentrations of the enzyme, the LDH levels in the blood serum will be elevated following cellular damage. The diagnostic value of LDH isoenzymes in liver diseases, however, should be assessed in respect to the animal species. Some examples of elevation in serum LDH levels in liver and skeletal muscle disorders are listed in Table 4.

Table 4. Some causes of increased LDH activity in serum (according to Bogin, 1994; Stockham & Scott, 2002)

Hepatocyte damage	
Degenerative	hypoxia due to anemia or congestion, cholelithiasis
Metabolic	lipidosis or fat cow syndrome, diabetes mellitus, equine hyperlipidemia
Neoplastic	lymphoma and metastatic neoplasia, hepatocellular carcinoma
Inflammatory infectious	bacterial and necrotic hepatitis, cholangiohepatitis, hepatic abscess
Inflammatory non-infectious	Theiler's disease, chronic hepatitis, cirrhosis
Toxic	iron toxicity, alkaloid containing plants, aflatoxins
Muscle damage – skeletal or cardiac muscle (mostly skeletal)	
Degenerative	hypoxia due to exertion or seisures, exertional rhabdomyolysis, saddle thrombus
Metabolic and nutritional	feline hyperthyroidism, vitamin E or selenium deficiency
Neoplastic	metastatic neoplasia
Inherited	musculodystrophy, hyperkalemic myopathy
Inflammatory	myositis (*Neospora, Toxoplasma*), bacteria or other agents
Toxic	monensin, castor bean, gossypol
Traumatic	intramuscular injection, hit-by-car, recumbency, seisures, exertion

Determination of serum LDH isoenzymes in combination with creatine kinase or other biochemical markers has become as well established procedure for helping in diagnosis of acute myocardial infarction in humans (Maekawa, 1988) with increased both LDH_1 and LDH_2 activities. In animals, the acute myocardial infarction rarely occurs making the test limited in veterinary medicine practice of heart attack, and, on the other hand useful in research (Preus et al., 1988; Aguilera-Tejero, 1989).

Relatively specific field of LDH activity determination in veterinary medicine practice is connected with the problem of mastitis in dairy cows. The increased level of total LDH in dairy cow milk serum was suggested to be a sensitive indicator of the udder epithelial cell damage and elevated numbers of leukocytes (Bogin et al., 1977; Hiss et al., 2007; Yang et al., 2011) with a higher proportion of LDH_4 and LDH_5.

Increase in LDH level has been detected in humans with various cancers and was proposed as a useful and inexpensive prognostic biomarker in patients with acute leukemia (Hu et al., 2016), metastatic carcinomas (Petrelli et al., 2015) and non-Hodgkin´s lymphoma (Huijgen et al., 1997). In veterinary medicine an increase in the value of serum LDH has been reported in dogs affected by leukemia (Leifer & Matus, 1986) and in cats with large granular lymphoproliferative disorders (Buracco et. al., 1992). An increase in LDH_3 was found as a possible indicator of latent canine lymphoma. (Bezzecchi et al., 1979; Abate et al., 1997; Zanatta et al. 2003).

Serious problem in veterinary medicine are respiratory diseases in animals. According to the present knowledge, LDH is one of the perspective biochemical parameters indicating the lung damage (Drent et al., 1996) with the pleural effusion and bronchoalveolar lavage fluid as the sources of the enzyme. Measurement of the total LDH activity and the profile of its isoenzymes in lung lesion from lambs suffering from acute as well as chronic pneumonia showed increased LDH_4 and LDH_5 activity (Milney & Doxey, 1984). In humans with pulmonary involvement including various carcinomas, however, elevations in serum LDH_3 occur most frequently (Johnson-Davis & McMillin, 2010). Increased LDH activity with remarkable elevation of the LDH_1 fraction in serum has been observed in calves with clinical signs of lower respiratory tract diseases (Nagy et al., 2013). This finding did not correlate with the changes in the LDH isoenzyme pattern found in humans and various animal species. The reason of this discrepancy could be connected with the presence of polymorphonuclear cells and alveolaer macrophages as infiltrating cells (Arai et al., 2003; Tanaka et al., 2006).

Abnormal pattern with increased LDH_5 activity in serum was observed in pigs and cattle under the physical stress (Moss & McMurray, 1979; Weeding et al., 1993; Heinová et al., 1994; Heinová & Vaško, 2007). The change in the LDH isoenzyme pattern in these animals was characterized by a presence of an extraordinarily intensive diffuse zone in the area of the LDH_4 cathodic site. Similar patterns were observed in some human serum/plasma where LDH zones were identified as macroglobular complexes of LDH isoenzymes with immunoglobulins (Biewenga & Feltkamp, 1975; Tozuka et al., 1996; Perry et al., 1997) and were denoted LDH-macroenzymes. However, their pathological significance is not clear since they have been observed in patients suffering from different diseases as well as in clinically healthy individuals.

Conclusion

Determination of total LDH activity and its isoenzyme pattern in serum can contribute to the diagnosis of diseases which are linked to the damage of some organs. The identification of isoenzymes could increase the diagnostic value of LDH. However, isoenzyme analysis requires special assays, and species variations in tissue distributions make interpretations rather demanding. As the LDH activity provides additional information it can be considered as a valuable adjunct indicator in laboratory diagnostics of animal diseases.

References

Abate, O., Zanatta, R., Buracco, P., Miniscalco, B. & Bonioli, A. (1997). Significato diagnostico degli isoenzimi della lattato deidrogenasi sierica (LDH) nel linfoma canino. *Atti S.I.S. Vet., 51*, 579-580.

Aguilera-Tejero, E., Mayer-Valor, R. & Gómez-Cárdenas, G. (1989). Spectrophotometric method for differentiation of cardiac and hepatic lactate dehydrogenase activities in dogs. *Am. J. Vet. Res., 7,* 1128-1130.

Anderson, M. G. (1976). The effect of exercise on the lactic dehydrogenase and creatine kinase isoenzyme composition of horse serum. *Res. Vet. Sci., 20,* 191-196.

Appella, E. & Markert, C. L. (1961). Dissociation of lactate dehydrogenase into subunits with guanidine hydrochloride. *Biochem. Biophys. Res. Commun., 6,* 171-176.

Babaei, H., Mansouri-Najand, L., Molaei, M. M., Kheradmand, A. & Sharifan, M. (2007). Assessment of lactate dehydrogenase, alkaline phosphatase and aspartate aminotransferase activities in cow's milk as an indicator of subclinical mastitis. *Vet. Res. Commun., 31*, 419-425.

Battelino, L. J. & Blanco, A. (1970). Catalytic properties of the lactate dehydrogenase isoenzyme "X" from mouse testis. *J. Exp. Zool., 174,* 173-186.

Beatty, E. M. & Doxey, D. L. (1983). Lactate dehydrogenase and creatine kinase isoenzyme levels in the tissue and serum of normal lambs. *Res. Vet. Sci, 35,* 325-330.

Bezzecchi, G., Zanetti, G., Ubaldi, A. & Corbella, E. (1979). Isoenzimi dell'LDH nie cani di razze giganti. *Clin. Vet., 102*, 139-144.

Biewenga, J. & Feltkamp, T. E. W. (1975). Lactate dehydrogenase (LDH) – IgG3 immunoglobulin complexes in human serum. *Clin. Chim. Acta, 64,* 101-116.

Bogin, E., Ziv, G., Avidar, J., Rivetz, B., Gordin, S. & Saran, A. (1977). Distribution of lactate dehydrogenase isoenzymes in normal and inflamed bovine udders and milk. *Res. Vet. Sci., 22*, 198-200.

Bogin, E. (1994). *Handbook for Veterinary Clinical Biochemistry*. New York, USA. Kodak Publ.

Bonny, C., Cooker, L. D. & Goldberg, E. (1988). Deoxyrlbonucleic acid-protein interactions and expression of the human testis-specific lactate dehydrogenase protomer: transcription factor Sp1 plays a major role. *Biol. Reprod., 58,* 754-759.

Boyd, J. W. (1962). The comparative activity of some enzymes in sheep, cattle and rabbits – normal serum and tissue levels and changes during experimental liver necrosis. *Res. Vet. Sci., 3,* 256-268.

Brooks, G. A., Dubouchaud, H., Brown, M., Sicurello, J. P. & Butz, C.E. (1999). Role of mitochondrial lactate dehydrogenase and lactate oxidation in the intracellular lactate shuttle. *Proc. Natl. Acad. Sci. USA, 96,* 1129-1134.

Buracco, P., Guglielmino, R., Abate, O., Bocchini, V., Cornaglia, E., De Nicola, D. B., Cilli, M. & Ponzio, P. (1992). Large granular lymphoma in an FIV-positive and FeLV-negative cat. *J. Small Anim. Pract., 33*, 279-284.

Cahn, R. D., Kaplan, N. O., Levine, L. & Zwilling. E. (1962). Nature and development of lactic dehydrogenases. The two major types of this enzyme form molecular hybrids which change in makeup during development. *Science, 136,* 962-969.

Cardinett III, G. H. (1997). Skeletal muscle function. In Kaneko, J.J., Harvey, J. W. & Bruss, M. L. (Eds.), *Clinical Biochemistry of Domestic Animals* (Fifth Edition pp. 1-932). San Diego, New York, USA: Academic Press.

Chaplin, S. B., Munson, M. M. & Knuth, S. T. (1997). The effect of exercise and restrain on pectoral muscle metabolism in pigeons. *J. Comp. Physiol. B., 167,* 197-203.

Davies, D. H., Long, D. L., McCarthy, A. R. & Herceg, M. (1986). The effect of parainfluenza virus type 3 on the phagocytic cell response of the ovine lung to *Pasteurella haemolytica. Vet. Microbiol., 11*, 125-144.

Dawson, D. M., Goodfriend, T. L. & Kaplan, N. O. (1964). Lactic dehydrogenases: function of the two types. *Science, 143,* 929-933.

Elustondo, P. A., White, A. E., Hughes, M. E., Brebner, K., Pavlov, E. & Kane, D. A. (2013). Physical and functional association of lactate dehydrogenase (LDH) with skeletal muscle mitochondria. *J. Biol. Chem., 288,* 25309-25317.

Fondy, T. P., Pesce, A., Freedberg, I., Stolzenbach, F. & Kaplan, N. O. (1964). The comparative enzymology of lactic dehydrogenase. II. Properties of the crystalline HM_3 hybrid from chicken muscle and H_2M_2 hybrid and H_4 enzyme from chicken liver. *Biochemistry, 4,* 522-530.

Fondy, T. P & Kaplan, N. O. (1965). Structural and functional properties of the H and M subunits of lactic dehydrogenases. *Ann. NY Acad. Sci., 119,* 888-903.

Frey, J., Haldimann, A., Kobisch, M & Nicolet, J. (1994). Immune response against the L-lactate dehydrogenase of Mycoplasma hyopneumoniae in enzootic pneumonia of swine. *Microb. Pathog., 17,* 313-322.

Goldberg, E., Eddy, E. M., Duan, Ch. & Odet, F. (2010). LDHC the ultimate testis specific gene. *J. Androl., 31,* 86-94.

Hedges, S. B. (1994). Molecular evidence for the origin of birds. *Proc. Natl. Acad. Sci USA, 91,* 2621-2624.

Heinová, D., Blahovec, J. & Kováč, G. (1994). New aspects of lactate dehydrogenase isoenzyme pattern in the serum of pigs at slaughter. *Vet. Med.-Czech, 39,* 287-296.

Heinová, D., Blahovec, J. & Rosival, I. (1996). Lactate dehydrogenase isoenzyme patterns in bird, carp and mammalian sera. *Eur. J. Clin. Chem. Clin. Biochem., 34,* 91-95.

Heinová, D., Rosival, I., Avidar, Y. & Bogin, E. (1999). Lactate dehydrogenase isoenzyme distribution and patterns in chicken organs. *Res. Vet. Sci., 67,* 309-312.

Heinová, D. & Vaško, L. (2007). Lactate dehydrogenase isoenzyme pattern in serum of cattle at slaughter. *Folia Veterinaria, 51,* 14-18.

Heinová, D., Kostecká, Z. & Csank, T. (2016). Separation of turkey lactate dehydrogenase isoenzymes using isoelectric focusing technique. *Electrophoresis, 37,* 335-338.

Hiss, S., Mueller, U., Neu-Zahren, A. & Sauerwein, H. (2007). Haptoglobin and lactate dehydrogenase measurements in milk for the identification of subclinically diseased udder quarters. *Vet. Med., 52*, 245-252.

Holland, L. Z., McFall-Ngai, M. & Somers, G. N. (1997). Evolution of lactate dehydrogenase-A homologs of barracuda fishes (genus Sphyranea) from different thermal environments: differences in kinetic properties and thermal stability are due to amino-acid substitutions outside the active site. *Biochemistry, 36,* 3207-3215.

Holmes, R. S. (1972). Evolution of lactate dehydrogenase genes. *FEBS Lett., 28,* 51-55.

Hu, W., Wang, X. & Yang, R. (2016). Evaluation of D-dimer and lactate dehydrogenase plasma levels in patients with relapsed acute leukemia. *Oncol. Lett., 12,* 591-596.

Huijgen, H., Sanders, G. T., Koster, R. N., Vreeken, J. & Bossuyt, P. M. (1997). Clinical value of lactate dehydrogenase in serum: a quantitative review. *Eur. J. Clin. Chem. Clin. Biochem., 35,* 569-579.

Kopperschlager, G. & Kirchberger, J. (1996). Methods for the separation of lactate dehydrogenases and clinical significance of the enzyme. *J. Chromatogr. B, Biomed. Appl., 684,* 25-49.

Leifer, C. E. & Matus, R. E. (1986). Chronic lymphocitic leukemia in the dog: 22 cases (1974-1984). *J. Am. Vet. Med. Assoc., 189*, 214-217.

Li, S. S. L., Fitch, W. M., Pan, Y. C. E. & Sharief, F. S. (1983). Evolutionary realationship of vertebrate lactate dehydrogenase genes A (muscle), B (heart) and C (testis). *J. Biol. Chem., 258,* 7029-7032.

Li, Y. J., Tsoi, S. C., Mannen, H. & Li, S. S. (2002). Phylogenetic analysis of vertebrate lactate dehydrogenase (LDH) multigene families. *J. Mol. Evol., 54,* 614-624.

Maekawa, M. (1988). Review. Lactate dehydrogenase isoenzymes. *J. Chromatogr., 429,* 373-398.

Mannen, H., Tsoi, S. C. Krushkal, J. S., Li, W. H. & Li, S. S. (1997). The cDNA cloning and molecular evolution of reptile and pigeon lactate dehydrogenase isoenzymes. *Mol. Biol. Evol., 11,* 1081-1087.

Markert, C. L., Shaklee, J. B. & Whitt, G. S. (1975). Evolution of gene. *Science, 189,* 102-114.

Milne, E. M. & Doxey, D. L. (1987). Lactate dehydrogenase and its isoenzymes in the tissues and sera of clinically normal dogs. *Res. Vet. Sci., 43,* 222-224.

Moss, B. W. & McMurray, C. H. (1979). The effect of duration and type of stress on some serum enzyme levels in pigs. *Res. Vet. Sci., 26,* 1-6.

Nagy, O., Tóthová, Cs., Seidel, H., Paulíková, I. & Kováč, G. (2011). Serum lactate dehydrogenase and its isoenzyme patterns in calves suffering from respiratory diseases. *Veterinarska stanica, 42 – Supplement 2,* 350-356.

Nečas, A. & Sedláková, D. (1999). Changes in the creatine kinase and lactate dehydrogenase activities in cerebrospinal fluid of dogs with thoracolumbar disc diseases. *Acta Vet. Brno, 68,* 111-120.

Perry, C., Peretz, H., Graf, E., Ben-Tal, O. & Eldor, A. (1997). Highly elevated lactate dehydrogenase level in a healthy individual: a case of macro-LDH. *Am. J. Hematol., 55,* 39-40.

Petrelli, F., Cabiddu, M., Coinu, A., Borbonovo, K., Ghiraldi, M., Lonati, V. & Barni, S. (2015). Prognostic role of lactate dehydrogenase in solid tumors: A systematic review and meta-analysis of 76 studies. Review article. *Acta Oncol., 54,* 961-970.

Plagemann, P. G. W., Gregory, K. F. & Wróblewski, F. (1960a). The electrophoretically distinct forms of mammalian lactic dehydrogenase. I. Distribution of lactic dehydrogenase in rabbit and human tissues. *J. Biol. Chem., 8,* 2282-2287.

Plagemann, P. G. W., Gregory, K. F. & Wróblewski, F. (1960b). The electrophoretically distinct forms of mammalian lactic dehydrogenase. II. Properties and interrelationships of rabbit and human lactic dehydrogenase isoenzymes. *J. Biol. Chem., 8,* 2288-2293.

Plummer, D. T. & Wilkinson, J. H. (1962). Organ specificity and lactate dehydrogenase activity. 2. Some properties of human-heart and liver preparations. *Biochem. J., 87,* 423-429.

Preus, M., Bhargava, A. S., Khater, A. E. R. & Günzel, P. (1988). Diagnostic value of serum creatine kinase and lactate dehydrogenase isoenzyme determination for monitoring early cardiac damage in rats. *Toxicol. Lett., 42,* 225-233.

Preus, M., Karste, B. & Bhargava, A. S. (1989). Seum isoenzyme pattern of creatine kinase and lactate dehydrogenase in various animal species. *J. Clin. Chem. Clin. Biochem., 27,* 787-790.

Quattro, J. M., Woods, H. A., & Powers, D. A. (1993). Sequence analysis of teleost retina specific lactate dehydrogenase C: evolutionary implications for the vertebrate lactate dehydrogenase gene family. *Proc. Natl. Acad. Sci. USA, 90,* 242-246.

Reinhold, P., Müller, G., Kreutzer, B., Gerischer, A. & Putsche, R. (1992). Diagnostische Aussagefähigkeit biochemischer und zytologischer Parameter in der Lungenspülflüsigkeit gesunder und pneumoniekranker Kälber. *J. Vet. Med. A, 39,* 404-418.

Roussel, J. D. & Stallcup, O. T. (1967). Distribution of lactic dehydrogenase and transaminase in the genital tissues of Holstein-Friesian bulls. *J. Diary Sci., 50,* 8, 1306-1309.

Salehi-Ashtiani, K. & Goldberg, E. (1993). Differences in regulation of testis specific lactate dehydrogenase in rat and mouse occur at multiple levels. *Mol. Reprod. Dev., 35,* 1-7.

Schmökel, H., Messmer, M., Lutz, H. (2001). Messung der Aktivitäten der alkalischen Phosphatase, Aspartataminotransferase und Laktatdehydrogenase in der Synovialflüssigkeit gesunder und veränderter Gelenke beim Hund. *Wien. Tierärztl. Mschr., 88,* 118-121.

Schurr, A. (2014). Cerebral glycolysis: a century of persistent misunderstanding and misconception. Review Article. *Front. Neurosci., 8,* 1-18.

Singh, R. P., Sastry, K. V. H., Agarwal, R., Pandey, N. K., Saxena, V. K. & Mohan, J. (2012). Molecular and kinetic characterization of lactate dehydrogenase enzyme in the heart and breast muscle of Japanese quail (Cortunix japonica). *Turk. J. Vet. Anim. Sci., 36,* 628-634.

Stockham, S. L. & Scott, M. A. (2002). *Fundamentals of Veterinary Clinical Pathology*. Blackwell Publishing Company, Iowa State Press.

Thortton, J. R. & Lohni, M. D. (1979). Tissue and plasma activity of lactic dehydrogenase and creatine kinase in the horse. *Equine Vet. J., 11,* 235-238.

Tozuka, M., Hidaka, H., Okumura, N., Ichikawa, T., Furihata, K. & Katsuyama, T. (1996). A case of immunoglobulin A – lambda conjugated with lactate dehydrogenase – 5 isoenzyme causing an extreamly high enzyme activity in serum. *Clin. Chem., 42,* 1288-1290.

Tsoi, S. C. M. & Li, S. S. L. (1994). The nucleotide and deduced amino-acid sequences of a cDNA encoding lactate dehydrogenase from

Caenorhabditis elegans: the evolutionary relationship of lactate dehydrogenases from mammals, birds, amphibian, fish, nematode, plants, bacteria, mycoplasm, and plasmodium. *Biochem. Biophys. Res. Commun., 205,* 558-564.

Tsuji, S., Quereshi, M. A., Hou, E. W., Fitch, W. M. & Li, S. S. L. (1994). Evolutionary relationships of lactate dehydrogenases (LDHs) from mammals, birds, an amphibian, fish, barley, and bacteria: LDH cDNA sequences from xenopus, pig, and rat. *Proc. Natl. Acad. Sci. USA, 91,* 9392-9396.

Tzouvelekis, A., Kouliatsis, G., Anevlavis, S. & Bouros, D. (2005). Serum biomarkers in interstitial lung diseases. *Resp. Res., 6*, 78. Available from: http:// respiratory-research.com/content/6/78.

Weeding, C. M., Hunter, E. J., Guise, H. J. & Penny, R. H. C. (1993). Effects of abatoir and slaughter handling systems on stress indicators in pig blood. *Vet. Rec., 133,* 10-13.

Weiss, D. J., Bauer, M. C., Whiteley, L. O., Maheswaran, S. K. & Ames, T. R. (1991). Changes in blood and bronchoalveolar lavage fluid components in calves with experimentally induced pneumonic pasteurellosis. *Am. J. Vet. Res., 52,* 337-344.

Wilson, A. C., Cahn, R. D. & Kaplan, N. O. (1963). Functions of the two forms of lactic dehydrogenase in the breast muscle of birds. *Nature, 197,* 331-334.

Wu, K. C., Chan, K., Lee, C. Y. & Lau, Y. F. (1987). Molecular isolation and sequence determination of the cDNA for the mouse sperm-specific lactate dehydrogenase – X gene. *Biochem. Biophys. Res. Common., 146,* 964-970.

Yang, F. L., Li, X. S., He, B. X., Yang, X. L., Li, G. H., Liu, P., Huang, Q. H., Pan, X. M. & Li, J. (2011). Malondialdehyde level and some enzymatic activities in subclinical mastitis milk. *Afr. J. Biotechnol., 10*, 5534-5538.

Yasuda, J., Tateyama, K., Syuto, B. &Too, K. (1990). Lactate dehydrogenase and creatine phosphokinase isoenzymes in tissues of laboratory animals. *Jpn. J. Vet. Res., 38,* 19-29.

Zanatta, R., Abate, O., D'Angelo, A., Miniscalco, B. & Mannelli, A. (2003). Diagnostic and prognostic value of serum lactate dehydrogenase (LDH) and LDH isoenzymes in canine lymphoma. *Vet. Res. Commun. 27 Suppl.,* 449-452.

Zinkham, W. H. & Isensee, H. (1972). Genetic control of lactate dehydrogenase synthesis in the somatic and gametic tissues of pigeon. *Johns. Hopkins. Med. J., 130,* 11-25.

BIOGRAPHICAL SKETCHES

Dagmar Heinová

Affiliation: Department of Chemistry, Biochemistry and Biophysics, University of Veterinary Medicine and Pharmacy in Košice, Slovakia

Education: DVM (1981) – University of Veterinary Medicine and Pharmacy in Košice, Slovakia; PhD thesis (1987) – Hydrolysis of free and acylated derivatives of p-nitroanilide by chymotrypsin and pepsin

Business Address: Komenského 73, 041 81 Košice, Slovak Republic

Research and Professional Experience: Since 1981, proteolytic enzymes, lactate dehydrogenase isoenzymes

Professional Appointments: Head of subject: Clinical Biochemistry for study programme General Veterinary Medicine (2008)

Honors: Associated Professor since 2005

Publications:

Heinová, D., Kostecká, Z., Csank, T. (2016). Separation of turkey lactate dehydrogenase isoenzymes using isoelectric focusing technique. *Electrophoresis, 37,* 335-338.

Dobríková, I., Kostecká, Z., Heinová, D. (2013). Influence of soybean diet on aminopeptidase activity of housefly larvae. *Folia veterinaria, 57,* 104-106.

Korének, M., Koréneková, B., Jalč, P., Mačanga, J., Heinová, D., Brenesselová, M. (2012). The changes in lactate dehydrogenase activity during ripening process of killed and shot wild rabbit´s meat. *Folia veterinaria, 56,* 36-38.

Zuzana Kostecká

Affiliation: Department of Chemistry, Biochemistry and Biophysics, University of Veterinary Medicine and Pharmacy in Košice, Slovakia

Education: DVM (1991) – University of Veterinary Medicine and Pharmacy in Košice, Slovakia; PhD. thesis (2003) Isolation of peptide growth factors from animal body fluids

Business Address: Komenského 73, 041 81 Košice, Slovak Republic

Research and Professional Experience: since 1991; peptide growth factors; peptidases in larvae of housefly

Professional Appointments:

Head of Department of Chemistry, Biochemistry and Biophysics (2011)

Head of Institute of Biochemistry (2007)

Head of subjects: Biochemistry for study programme General Veterinary Medicine (GVM) (2007)

Biochemistry for study programme Pharmacy (2007)

Pathological Biochemistry for study programme GVM (2006)

Clinical and Pathological Biochemistry for study programme Pharmacy (2014)

Honors: Associated Professor (2010)

Publications:

Macko, J., Oriňáková, R., Oriňák, A., Kovaľ, K., Kupková, M., Erdélyi, B., Kostecká, Z., Smith, R. M. (2016). Hydrophobicity of silver surfaces with microparticle geometry. *Applied Surface Science, 387,* 1195-1201.

Heinová, D., Kostecká, Z., Csank, T. (2016). Separation of turkey lactate dehydrogenase isoenzymes using isoelectric focusing technique. *Electrophoresis, 37,* 335-338.

Valko-Rokytovská, M., Bruchatá, K., Šimková, J., Milkovičová, M., Kostecká, Z. (2016). Current trends in the treatment of malignant melanoma. Minireview. *Neoplasma, 63,* 3, 333-341.

Konvičná, J., Vargová, M., Paulíková, I., Kováč, G., Kostecká, Z. (2015). Oxidative stress and antioxidant status in dairy cows during prepartal and postpartal periods. *Acta Veterinaria Brno, 84,* 133-140.

Dobríková, I., Kostecká, Z., Heinová, D. (2013). Influence of soybean diet on aminopeptidase activity of housefly larvae. *Folia veterinaria, 57*, 104-106.

Oskar Nagy

Affiliation: Clinic of Ruminants, University of Veterinary Medicine and Pharmacy in Košice, Slovakia

Education: DVM (1985) - University of Veterinary Medicine and Pharmacy in Košice; PhD. thesis (1999) – Acid-base balance and blood gases in calves in relation to health and diseases

Business Address: Komenského 73, 041 81 Košice, Slovak Republic

Research and Professional Experience: diseases of ruminants, internal diseases, blood biochemistry, clinical pathology, acid- base, blood gases, serum proteins

Professional Appointments:

Member of European College of Bovine Health Management since 2007

Member of Association of Slovak Food Animal Veterinary Surgeons since 2011

Head of subject: Diagnostic of disorders of internal environment of animals

Vice-rector of UVMP (2011)

Honors: Associated Professor since 2007

Publications:

Šoltésová, H., Nagy, O., Tóthová, Cs., Paulíková, I., Seidel, H. (2015). Blood gases, acid-base status and plasma lactate concentrations in calves with respiratory diseases. *Acta Veterinaria Beograd., 65,* 111-124.

Tóthová, Cs., Nagy, O., Nagyová, V., Kováč, G. (2015). Changes in the concentrations of acute phase proteins in calves during the first month of life. *Acta Veterinaria Beograd*, 65, 260-270.

Nagy, O., Tóthová, Cs., Nagyová, V., Kováč, G. (2015). Comparison of serum protein electrophoretic pattern in cows and small ruminants. *Acta veterinaria Brno,* 84, 187-195.

Šoltésová, H., Nagyová, V., Tóthová, Cs., Nagy, O. (2015). Haematological and blood biochemical alterations associated with respiratory disease in calves. *Acta veterinaria Brno,* 84, 249-256.

Nagyová, V., Šoltésová, H., Tóthová, Cs., Nagy, O. (2015). The impact of respiratory disease and diarrhoea on protein profiles in calves. *Folia veterinaria, 59,* 5-11.

Tóthová, Cs., Nagy, O., Kováč, G. (2015). Comparison of the results pf some biochemical variables obtained by portable chemistry analyzer and standard laboratory methods. *Folia veterinaria, 59,* 30-38.

Tóthová, Cs., Nagy, O., Kováč, G. (2014). Major bovine acute phase proteins in relation to sample storage temperature and time. *Cardiovascular Disease II.* 1. vyd. - Hong Kong : iConcept Press, 560.

Tóthová, Cs., Nagy, O., Kováč, G. (2014). Changes in the concentrations of selected acute phase proteins and variables of energetic profile in dairy cows after parturition. *Journal of Applied Animal Research*, 42, 278-283.

Nagy, O., Tóthová, Cs., Kováč, G. (2014). Age-related changes in the concentrations of serum proteins in calves. *Journal of Applied Animal Research*, 42, 451-458.

Tóthová, Cs., Nagy, O., Kováč, G. (2014). Acute phase proteins and their use in the diagnosis of diseases in ruminants: a review. *Veterinární medicína*, 59, 163-180.

Tóthová, Cs., Nagy, O., Kováč, G. (2014) Relationship between some variables of protein profile and indicators of lipomobilization in dairy cows after calving. *Archiv fur Tierzucht.*, 57, 1-9.

Nagy, O., Tóthová, Cs., Nagyová, V., Kováč, G., Pošivák, J. (2014). Changes in the serum protein electrophoretic pattern in lambs during the first month of life. *Acta veterinaria Brno,* 83, 187-193.

Tóthová, Cs., Nagy, O., Kováč, G. (2014). Changes in the concentrations of serum protein fractions in calves with age and nutrition. *Italian Journal of Animal Science,* 13, 107-111.

Lipták, T., Capík, I., Ledecký, V., Nagy, O., Kuriocvá, M., Tóthová, Cs., Maďari, A., Farbáková, J., Petrovič, V., Horňák, S. (2014). Cardiorespiratory and hemodynamic effects of medetomidine or xylazine with atropine and diazepam premedication for total intravenous anesthesia induced and maintained with propofol/fentanyl in dogs undergoing surgery. *Acta Veterinaria Beograd,* 64, 466-476.

Nagy, O., Paulíková, I., Seidel, H., Tóthová, Cs., Šoltésová, H. (2014). Clinical signs, blood gases and acid-base balance after nitrate administration in calves. *Comparative Clinical Pathology,* 23, 647-652.

Nagy, O., Tóthová, Cs., Seidel, H., Paulíková, I., Kováč, G. (2013). The effect of respiratory diseases on serum lactate dehydrogenase and its isoenzyme patterns in calves. *Polish Journal of Veterinary Sciences,* 16, 211-218.

Tóthová, Cs., Nagy, O., Kováč, G. (2013). The serum protein electrophoretic pattern and acute phase protein concentrations in calves with chronic respiratory diseases. *Polish Journal of Veterinary Sciences,* 16, 385-386.

Tóthová, Cs., Nagy, O., Kováč, G. (2013). The serum protein electrophoretic pattern and acute phase proteins concentrations in calves with chronic respiratory diseases. *Acta Veterinaria Beograd,* 63, 473-486.

Tóthová, Cs., Nagy, O., Seidel, H., Kováč, G. (2013). Serum protein electrophoretic pattern in clinically healthy calves and cows determined by agarose gel electrophoresis. *Comparative Clinical Pathology,* 22, 15-20.

Kováč, G., Tóthová, Cs., Nagy, O. (2013). The use of acute phase proteins in laboratory diagnosis of mastitis. *Folia veterinaria,* 57, suppl. 1, 49.

In: Dehydrogenases
Editor: Margaret Collins
ISBN: 978-1-53610-550-6

Chapter 2

Renal Betaine Aldehyde Dehydrogenase Activity Is Modulated Synergistically by Urea, Sodium and Potassium Under Physiological Conditions

Hector G. Ayala-Castro, Ciria G. Figueroa-Soto and Elisa M. Valenzuela-Soto*
Coordinación de Ciencia de los Alimentos,
Centro de Investigación en Alimentación y Desarrollo A. C.,
Hermosillo, Sonora, México

Abstract

In kidney the role of urea is to modulate urine concentration, however, urea is a protein denaturant with a strong effect on several renal enzymes. The enzyme betaine aldehyde dehydrogenase catalyzes the synthesis of glycine betaine and is located in cortex and medulla cells of kidney, and is therefore, exposed to high urea concentrations. Betaine aldehyde dehydrogenase follows an ordered sequential kinetic mechanism with NAD^+ as the first substrate to bind to the enzyme. The

* Corresponding author: Tel: +52 (662) 289-2400, Fax: +52 (662) 280-0058; e-mail: elisa@ciad.mx.

effect of urea in the BADH kinetic parameters was analyzed under two ionic strength conditions. Enzyme activity was assayed at 0-2 M concentrations of urea and pH 7.0. The results showed that urea was a competitive inhibitor with respect to NAD^+ and non-competitive regarding the betaine aldehyde substrate. The competitive and non-competitive inhibition constants were calculated. Initial velocity patterns in presence of 150 mM NaCl or KCl showed that both salts increased the affinity of BADH for NAD^+. These results suggest that enzyme inhibition by urea is reversed by intracellular potassium concentrations, ensuring the synthesis of glycine betaine; an important osmolyte and osmoprotector in kidney. During the mechanism of urine concentration, the activity of betaine aldehyde dehydrogenase is maintained despite the high urea concentrations because of the hyperosmolarity required to produce concentrated urine.

Keywords: betaine aldehyde dehydrogenase, glycine betaine, inhibition, kidney, ionic strength, osmoregulation, urea

INTRODUCTION

In mammals, the most abundant protein catabolism waste product is nitrogen, which is excreted as urea in the urine. Urea concentration in the urine exceeds 1 M during antidiuresis [1]. Urea plays an important role during the urinary concentrating mechanism because the passive or active accumulation of urea in renal medullary interstitial fluid increases the maximum urine concentrations that can occur in the kidney [2-4]. However, urea destabilizes proteins, DNA and RNA and therefore, can also have deleterious effects [5-8]. In addition, urea inhibits important protein functions such as ligand binding [9].

There is ample literature describing the effect of urea on protein denaturation and structure at high concentrations, however there are few kinetic analyses of the reversible inhibition of enzyme activity by urea [10-12], and also, scarce kinetic studies about the effect of urea on renal enzymes [13-15]. Such studies enable to better understanding about the role of enzymes in an environment as adverse as that of the renal medulla.

Renal medulla cells are exposed to high concentrations of urea as well as levels of hypertonicity so extreme, that they would be lethal in general circulation. These cells adapt to their environment through the controlled accumulation of compatible organic osmolytes [16]. Mammalian medulla cells accumulate glycine betaine (GB) glycerophosphocholine, sorbitol and inositol as osmolytes to contend with this hypertonic environment [16-18].

The synthesis of GB by renal tubule cells protects against the osmotic force of urine as it becomes concentrated in the distal tubules and collecting ducts [17]. GB is synthesized from betaine aldehyde (BA) in a reaction catalyzed by betaine aldehyde dehydrogenase (BADH, EC 1.2.1.8 betaine aldehyde: NAD^+ oxidoreductase). This eznzyme is located in the renal cortex and medulla [19].

Porcine kidney BADH (pkBADH) belongs to the ALDH superfamily and is classified as ALDH class 9 [20-22]. This enzyme has been purified [23] and kinetically characterized [24]. BADH follows an iso bi bi ordered steady state mechanism: NAD^+ is the first substrate to bind the enzyme while NADH is the last product that is released from the enzyme complex [24]. Active pkBADH is a tetramer composed by 54 kDa identical subunits [24].

In a previous study, it was demonstrated that pkBADH requires ionic strength to maintain its tetrameric conformation; at low ionic strength the enzyme dissociates, forming dimers which are inactive and very stable [25]. Additionally, the enzyme requires the physiological ionic strength provided by a monovalent cation for maximum thermostability [26]. With the aim of studying the possible relationship between urea and BADH in kidney, earlier studies by our research team demonstrated that incubation of the pure enzyme with urea concentrations above 3 M resulted in inactivation of the pkBADH at pH 7.0 and 30 °C; but this inactivation was prevented by high ionic strength [27]. In this study we analyze the effect of variable urea concentrations on pkBADH kinetic constants by analyzing inhibition patterns and the role played by the monovalent cations sodium and potassium which are important salts regulating the urine concentrating mechanism.

Materials and Methods

Enzyme Purification

PkBADH was purified to homogeneity from porcine kidney using a previously reported procedure [23]. The pure enzyme was stored at -20 °C in 10 mM potassium phosphate buffer, pH 7.0, 1 mM EDTA, 10 mM β-mercaptoethanol, 0.2 M KCl and 10% glycerol. The enzyme (0.15 mg/mL) was dialyzed overnight at 4 °C against 10 mM Hepes-KOH buffer, pH 7.0, 1 mM EDTA, 10 mM β-mercaptoethanol to eliminate KCl. PkBADH activity was measured spectrophotometrically as reported elsewhere [23].

Initial Velocity Patterns

Steady-state initial velocity studies were carried out at 30 ºC, in 0.1 M Hepes-KOH buffer, pH 7.0, containing 1 mM EDTA in a final volume of 0.4 mL. The concentration range tested for NAD^+ was 0.025 to 0.5 mM, and for BA was 0.05 to 0.6 mM. The effect of NaCl and KCl was tested using 10 mM HEPES-KOH plus 0.15 M KCl or 0.15 M NaCl. Reaction rates were linear over the 3 min assay period. All assays were done at least twice and on two different days, with duplicate measurements each time, using an Ultrospec 4000 spectrophotometer. The reaction was initiated by adding the enzyme.

Inhibition Assays

For the inhibition patterns, the NAD^+ concentrations were 0.025, 0.05, 0.01, 0.25, and 0.5 mM, and the BA concentrations were 0.05, 0.1, 0.2, 0.4, and 0.6 mM. The urea concentrations tested were 0, 0.5, 1.0, 1.5, and 2.0 M. The effect of NaCl and KCl was tested using 10 mM HEPES-KOH plus 0.15 M KCl or 0.15 M NaCl. All assays were done at least twice and on two different days, with duplicate measurements each time, using an Ultrospec 4000 spectrophotometer.

Data Analysis

Kinetic data were analyzed using the non-linear regression program GraFit 7.0 (Erithacus Software). Initial velocity data were fitted to the Michaelis-Menten equation for a sequential mechanism (Eq. 1) where A is NAD^+ and B is betaine aldehyde, Km_a and Km_b are Michaelis-Menten constants, and K_{ia} is the dissociation constant of E-NAD^+. Equations (2) to (5) were used to determine the inhibition constant for urea according with each type of inhibition:

$$v = \frac{V[A][B]}{K_{mA}[B] + K_{mB}[A] + K_{ia}K_{mB} + [A][B]} \qquad \text{(Eq. 1)}$$

Competitive inhibition

$$v = \frac{V_{\max} * [NAD^+]}{K_m\left(1 + \frac{[Urea]}{K_{i(Urea)}} + [NAD^+]\right)} \qquad \text{(Eq. 2)}$$

Non-competitive inhibition

$$v = \frac{V_{\max} * [BA] * \frac{1}{1 + [Urea]/K_{i(Urea)}}}{K_m + [BA]} \qquad \text{(Eq. 3)}$$

Dixon competitive inhibition

$$v = \frac{K_m + [NAD^+]}{V_{\max} * [NAD^+]} + \frac{[Urea] * \left(\frac{K_{m(NAD^+)}}{K_{i(Urea)}}\right)}{V_{\max} * [NAD^+]} \qquad \text{(Eq. 4)}$$

Dixon non-competitive inhibition

$$v = \frac{1}{V_{\max}} * \left(1 + \frac{K_{m(BA)}}{[BA]}\right) * \left(1 + \frac{[Urea]}{K_i}\right) \qquad \text{(Eq. 5)}$$

RESULTS

PkBADH kinetic constants were previously reported, having been determined at pH 8.0, but in the present study the aim was to analyze the effect of urea, NaCl and KCl on BADH under kidney physiological conditions; therefore V_{max} and K_m values for BA and NAD^+ were calculated at pH 7.0 from initial velocity patterns. The 1/v data from initial velocity patterns were reploted versus $1/[NAD^+]$ or [BA] concentrations to calculate the kinetic constants (Figure 1A and B respectively). BADH V_{max} was 1.18 ± 0.026 μmol/min/mg, whereas K_{mNAD} and K_{mBA} were 58.7 ± 4.3 μM and 166 ± 18.4 μM, respectively (Table 1). PkBADH retained its affinity for its substrates at pH 7.0, because at pH 8.0 K_{mNAD} and K_{mBA} were 44.5 ± 8.2 μM and 145.5 ± 8.9 μM, respectively [24].

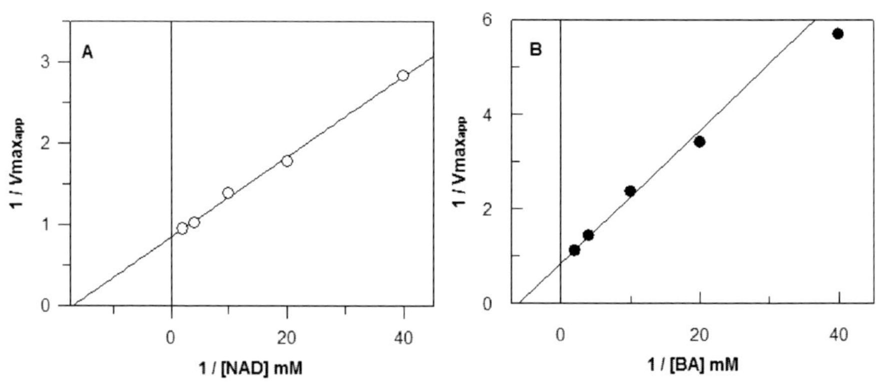

Figure 1. Double reciprocal plots of initial velocity versus substrate concentrations for porcine kidney betaine aldehyde dehydrogenase. Panel A: 1/Vmax apparent $[NAD^+]$. Panel B: 1/Vmax apparent *versus* [BA].

Table 1. Kinetic constants for porcine kidney BADH at pH 7.0 and in presence of KCl (0.15 M) and NaCl (0.15 M)

Constant	Value		
	None	KCl	NaCl
Vmax (μmoles/min mg)	1.18 ± 0.026	0.91 ± 0.04	0.9 ± 0.09
Km_{NAD} (μM)	58.7 ± 4.3	42.1 ± 4.9	44.4 ± 5.1
Km_{BA} (μM)	166 ± 18.4	337.3 ±32.8	335.5 ±33.7

The addition of KCl or NaCl provoked a slight decrease in V_{max} and increase in affinity constants of the pkBADH. The V_{max} was 0.91 ± 0.04 μmol/min/mg and 0.862 ±0.09 μmol/min/mg in presence of 0.15 M KCl and 0.15 mM NaCl respectively (Table 1), while K_{mNAD} and K_{mBA} were 42.1 ± 4.9

μM and 337 ± 32.8 μM, with 0.15 M KCl, and 49.4 ± 5.1 μM and 335.5 ± 33.7 μM with 0.15 M NaCl (Table 1).

The effect of urea on the pkBADH kinetic parameters was analyzed at a fixed and saturating BA concentration (0.6 mM) with NAD^+ concentration variable, the Lineweaver-Burk plots showed that urea is a competitive inhibitor with respect to NAD^+ (Figure 2A). The data points fit the competitive equation best (eq. 2), with an apparent V_{max} = 0.95 ± 0.02 μmol/min/mg and a K_{mNAD} = 22 ± 2.7 μM, while K_i was 330 ± 40 mM calculated using the Dixon competitive equation (eq. 4). The data for initial velocity *versus* urea concentration are shown in Figure 2B.

Under the NAD^+ saturating concentration (0.5 mM) the Lineweaver-Burk plot had a pattern of intersecting lines indicating non-competitive inhibition with respect to BA (Figure 3A). The data points fit the non-competitive equation best (eq. 3), with an apparent V_{max} = 1.5 ± 0.07 μmol/min/mg and K_{mBA} = 403 ± 30 μM (Table 2). A K_i of 1.91 ± 0.11 M was calculated using the Dixon equation (eq. 5) (Table 2). The Dixon plot of the initial velocity *versus* urea concentration data is given in the Figure 3B.

The effect of urea in presence of KCl (0.15 M) was analyzed and the Lineweaver-Burk plot of initial velocity versus NAD^+ showed that the type of inhibition was again, competitive (Figure 4A), with an apparent V_{max} of 0.91 ± 0.04 μmol/min/mg and a K_{mNAD} of 70 ± 9.6 μM. Changes in the initial velocity *versus* urea concentration are shown in the Figure 4B. The K_i calculated with the competitive Dixon equation (eq. 3) did not change in KCl presence (Table 2).

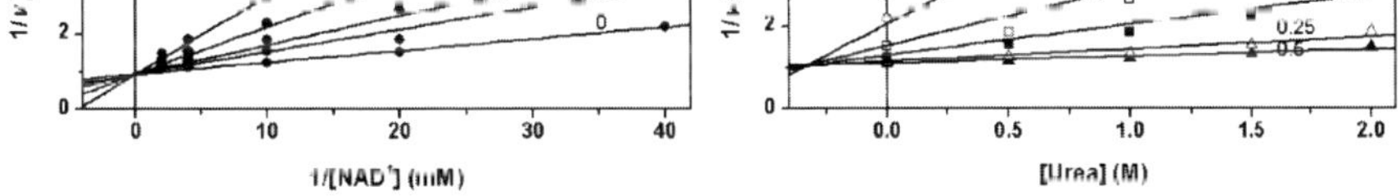

Figure 2. Inhibition of porcine kidney betaine aldehyde dehydrogenase by urea. Panel A: Lineweaver-Burk plot of initial velocity *versus* variable NAD^+ concentration. BA concentration was 0.6 mM and NAD^+ concentration was 0.5, 0.25, 0.1, 0.05, and 0.025 mM. Panel B: Dixon plot, $1/v$ against [urea], at NAD^+ concentrations showed.

On the other hand, the Lineweaver-Burk plot of initial velocity versus BA showed that the presence of 0.15 M KCl maintained a non-competitive inhibition (Figure 5A). An increase in the K_{mBA} (482 ± 71 μM) was observed, while the apparent V_{max} value did not change (Table 2). The K_i calculated with the Dixon non-competitive equation (eq. 4) showed that KCl decreased the K_i from 1.91 ± 0.11 M to 1.12 ± 0.08 M. Initial velocity and urea concentration data were used to obtain the Dixon plot shown in the Figure 5B.

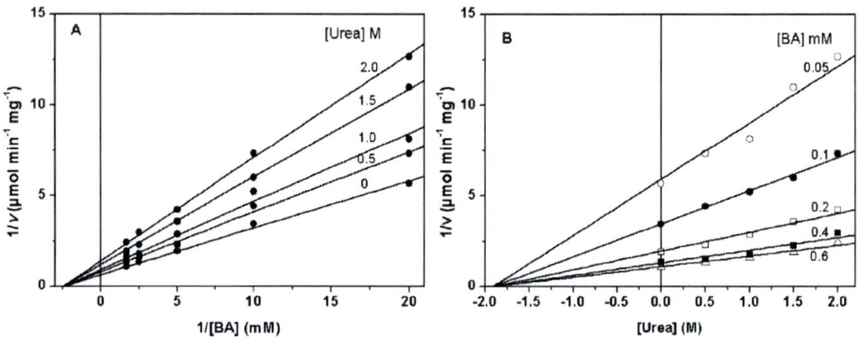

Figure 3. Inhibition of porcine kidney betaine aldehyde dehydrogenase by urea. Panel A: Lineweaver-Burk plot of initial velocity *versus* variable BA concentration. NAD^+ concentration was 0.5 mM and BA concentration was 0.05, 0.1, 0.2, 0.4 and 0.6mM. Panel B: Dixon plots, 1/*v* against [urea], at BA concentrations showed.

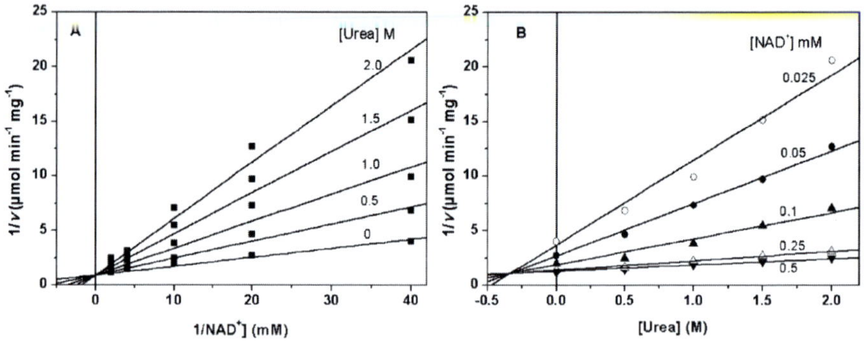

Figure 4. Effect of KCl on enzyme inhibition by urea. Panel A: Lineweaver-Burk plot of initial velocity *versus* variable NAD^+ concentration. BA concentration was 0.6 mM and NAD^+ concentrations were 0.025 mM, 0.05 mM, 0.1 mM, 0.25 mM and 0.5 mM. Panel B: Dixon plots, 1/*v* against [urea], at NAD^+ concentrations showed.

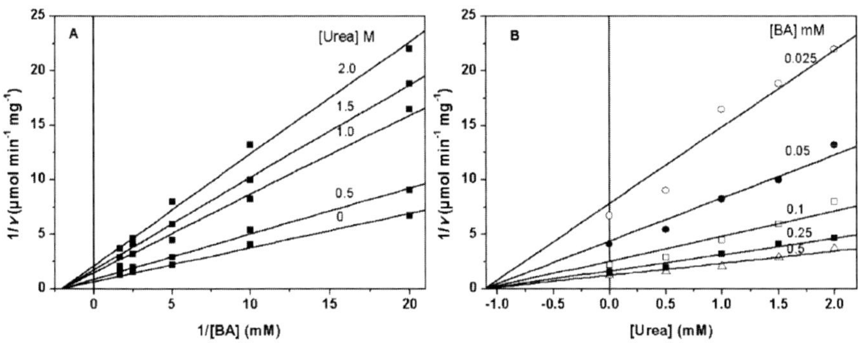

Figure 5. Effect KCl on enzyme inhibition by urea. Panel A: Lineweaver-Burk plot of initial velocity *versus* variable BA concentration. NAD^+ concentration was 0.5 mM and BA concentration was 0.05, 0.1, 0.2, 0.4 and 0.6mM. Panel B: Dixon plot, $1/v$ against [urea], at BA concentrations showed.

Table 2. Effect of urea on apparent kinetic constants of porcine kidney BADH at saturating substrate concentrations

Variable substrate	Fixed saturating substrate	V_{max} (μmol/min/mg)	K_m (μM)	K_i (M)	Kinetic pattern
NAD^+	BA	0.95 ± 0.02	22 ± 2.7	0.33 ± 0.04	Competitive
$NAD^+ + K^+$	BA	0.91 ± 0.03	70 ± 9.6	0.33 ± 0.04	Competitive
BA	NAD^+	1.5 ± 0.07	403 ± 32	1.91 ± 0.10	Non-competitive
$BA + K^+$	NAD^+	1.5 ± 0.12	182 ± 71	1.12 ± 0.08	Non-competitive

Discussion

During antidiuresis, renal medulla cells are exposed to hostile environment because high urea NaCl concentrations are needed to produce urine [28]. Urea may reach concentrations of 600 mM in the human kidney during antidiuresis, and can exceed 2 M in desert rodents [1, 29-31]. Medulla cells counteract high urea concentrations by accumulating osmolytes, including GB [16-18]. For rabbit renal medullary cells in tissue culture, GB content decreased in the presence of urea, suggesting that GB synthesis can be inhibited [29].

Therefore, it was important to determine changes in pkBADH activity and analyze the kinetic parameters under variable urea concentrations and the effect that KCl and NaCl can have on the enzyme activity, to explain how pkBADH can be active to synthesize GB in kidney. The data showed that potassium and sodium increased the enzyme affinity by NAD^+ (Table 1). PkBADH affinity changes by the substrate BA were found with both salts, the K_{mBA} was twice higher (Table 1). However, the K_{mBA} increase had low impact in V_{max} because it decreased 27% and 23% in presence of NaCl or KCl (respectively) (Table 1). It was interesting to find that sodium had little effect on the kinetic parameters of the enzyme, since some enzymes are inhibited by sodium [32, 33].

On the other hand, urea inhibited pkBADH activity which decreasing V_{max} 20%, interestingly the K_{mNAD} was 63% lower indicating an increase in enzyme affinity by NAD^+, a result similar to that found for aldose reductase [14]. Aldose reductase results were explained based on the chemical mechanism of the enzyme, in which the aldose reductase effects a conformational change in the active site to form the ENADPH complex necessary for the hydride transfer ($ENADP^+$) involved in the reduction of carbonyl [15]. In a previous study, we reported that the ligands of pkBADH induce slow conformational rearrangements at the active site, resulting in significantly decreased reactivity or accessibility of the essential Cys [34]. Similar results were found for BADHs from *P. aeruginosa* and amaranth leaves [35].

Therefore changes in K_{mNAD} do not necessarily reflect only changes in the affinity of the enzyme for the NAD^+, given that conformational changes are involved in the formation of the $pkBADHNAD^+$ complex. Likewise, urea can cause conformational changes at the pkBADH active site because it is a potential hydrogen bond acceptor and donor [36]. The amino acids that interact with NAD^+ at the active site are K180, G154, W156, S26, I326, S233, E391, G254 and W156, and these same residues may also be interacting with urea, and that interaction could be involved in enzyme inhibition.

When all the data are considered, an important question arises regarding how BADH can function adequately under the high urea conditions present in renal inner medulla cells. A possible explanation is that the medulla cells have saturating NAD^+ concentrations, since in rat renal cells NAD^+ concentration has been estimated to be 1 mM [37]. At that coenzyme concentration, the enzyme would be saturated with NAD^+ and the impact of urea inhibition would be reduced. In addition, it has been proposed that the effects of urea on proteins can be reduced by high NaCl concentrations [38], and these are present in medulla cells during antidiuresis.

During antidiuresis, potassium is the intracellular most abundant ion in medulla cells, and the inhibition data showed that urea plus potassium increased slightly the K_{mNAD} and K_{mBA}, but NAD^+ concentrations are within physiological range, which suggest that the enzyme can be active in spite of the high urea concentrations. Furthermore it has been demonstrated that pkBADH is localized both in cortex and medulla cells [19], therefore, it is feasible that GB is being synthesized by the enzyme in the cortex cells where the urea concentrations are lower and transported to the medulla cells.

To keep BADH activity in kidney is important to maintain adequate GB concentration since this compound has the property to be an osmolyte and osmoprotector [16, 18, 39-41]. In addition, GB regulates the levels and/or the activity of several enzymes and proteins involved in kidney metabolic routes. For example, in rat kidney Kim et al. found higher methionine adenosyltransferase (involved in sulfur amino acid metabolism) and in cysteine sulfinate decarboxylase (involved in taurine synthesis) [42]. Hyperuricemic mice treated with GB, also, presented higher urate transporter 1 and organic transporter 1, and a decrease in glucose transporter 9, all those proteins are involved in kidney urate excretion [43]. Therefore, it was proposed that GB reduces the uric acid in serum [43]. In apoptotic MDCK cells under hyperosmotic conditions supplemented with GB, the activity of caspase 3, 8 and 9 decreased in parallel to the reduction of cell damage [44].

In conclusion, high urea concentrations could cause pkBADH inhibition which can be counteracted by the presence of potassium which acts by increasing the affinity of the enzyme for the NAD^+substrate, that is the first substrate to bind to enzyme; and therefore, this explain the GB synthesis during antidiuresis in renal medulla cells.

ACKNOWLEDGMENTS

H. G. A. C. gratefully acknowledges a scholarship from CONACyT.

CONFLICTS OF INTEREST STATEMENT

The authors declare that there are no conflicts of interest.

REFERENCES

[1] Sands, J. M., Knepper, M. A., (1987). Urea permeability of mammalian inner medullary collecting duct system and papillary surface epithelium. *J. Clin. Invest.* 79, 138-147.

[2] Levinsky, N. G., Berliner, R. W. J., (1959). The role of urea in the urine concentrating mechanism. *Clin. Invest.* 38, 741-748.

[3] Beyer, Jr K. J., Gelarden, R. T., (1988). Active transport of urea by mammalian kidney. *Proc. Natl. Acad. Sci. USA* 85, 4030-4031.

[4] Yang, B., Bankier, L. (2005). Urea and urine concentrating ability: new insights from studies in mice. *Am. J. Physiol. Renal. Physiol.* 288, F881–F896.

[5] Bennion, B. J., Daggett, V., (2003). The molecular basis for the chemical denaturation of proteins by urea. *Proc. Natl. Acad. Sci. USA* 100, 5142–5147.

[6] Hong, J., Capp, M. W., Anderson, C. F., Saecker, R. M., Felitsky, D. J., Anderson, M. W., Record, M. T., (2004). Preferential interactions of glycine betaine and of urea with DNA: Implications for DNA hydration and for effects of these solutes on DNA stability. *Biochemistry* 43, 14744−14758.

[7] Guinn, E. J., Pegram, L. M., Capp, M. W., Pollock, M. N., Record, Jr M. T., (2011). Quantifying why urea is a protein denaturant, whereas glycine betaine is a protein stabilizer. *Proc. Natl. Acad. Sci. USA* 108, 16932-16937.

[8] Lambert, D., Draper, D. E., (2012). Denaturation of RNA secondary and tertiary structure by urea: Simple unfolded state models and free energy parameters account for measured m-values. *Biochemistry* 51, 9014-9026

[9] Yancey, P. H., (2005). Organic osmolytes as compatible, metabolic and counteracting cytoprotectants in high osmolarity and other stresses. *J. Exp. Biol.* 208, 2819-2830.

[10] Rajagopalan, K. V., Fridovich, I., Handler, P., (1961). Competitive inhibition of enzyme activity by urea. *J. Biol. Chem.* 236, 1059-1065.

[11] Fridovich, I., (1964). Competitive inhibition of xanthine oxidase by urea and guanidinium ion. *J. Biol. Chem.* 239, 3519-3521.

[12] Pocsi, I., Berki, A. C., Taylor, S. A., Price, R. G., (1993). Urea inhibition of urinary N-acetyl-beta-D-glucosaminidase is mixed- type and sensitive to changes of pH away from the optimum. *Clin. Chem.* 39, 1918-1919.

[13] Hand, S. C., Somero, G. N., (1982). Urea and methylamine effects on rabbit muscle phosphofructokinase. Catalytic stability and aggregation state as a function of pH and temperature. *J. Biol. Chem.* 257, 734-741.

[14] Burg, M. B., Peters, E. M., (1997). Urea and methylamines have similar effects on aldose reductase activity. *Am. J. Physiol. Renal Physiol.* 273, F1048-F1053.

[15] Burg, M. B., Peters, E. M., Bohren, K. M., Gabbay, K. H., (1999). Factors affecting counteraction by methylamines of urea effects on aldose reductase. *Proc. Natl. Acad. Sci. USA* 96, 6517-6522.

[16] Garcia-Perez, A., Burg, M. B., (1991). Renal medullary organic osmolytes. *Physiol. Rev.* 71, 1081-1115.

[17] Neuhofer, W., Beck, F., (2006). Survival in hostile environments: strategies of renal medullary cells. *Physiology* 21, 171-180.

[18] Burg, M. B., Ferraris, J. D., (2008). Intracellular organic osmolytes: function and regulation. *J. Biol. Chem.* 283, 7309-7313.

[19] Figueroa-Soto, C. G., López-Cervantes, G., Valenzuela-Soto, E. M., (1999). Immunolocalization of betaine aldehyde dehydrogenase in porcine kidney. *Biochem. Biophys. Res. Commun.* 258, 732-736.

[20] Vasiliou, V., Bairoch, A., Tipton, K. F., Nebert, D. W., (1999). Eukaryotic aldehyde dehydrogenase (ALDH) genes: human polymorphisms, and recommended nomenclature based on divergent evolution and chromosomal mapping. *Pharmacogenetics* 9, 421-434.

[21] Vasiliou, V., Thompson, D. C., Smith, C., Fujita, M., Chen, Y., (2013). Aldehyde dehydrogenases: From eye crystallins to metabolic disease and cancer stem cells, *Chem. Biol. Interact.* 202, 2-10.

[22] Julián-Sánchez, A., Riveros-Rosas, H., Martínez-Castilla, L. P., Velasco-García, R., Muñoz-Clares, R. A., (2007). Phylogenetic and structural relationships of the betaine aldehyde dehydrogenases. In: Weiner H, Plapp B, Lindahl R, Maser E, eds. Enzymology and Molecular Biology of Carbonyl Metabolism. West Lafayette: *Purdue University Press*, pp 64-76.

[23] Guzman-Partida, A. M., Valenzuela-Soto, E. M., (1998). Porcine kidney betaine aldehyde dehydrogenase: purification and properties. *Comp. Biochem. Physiol.* Part B 119, 485-491.

[24] Figueroa-Soto, C. G., Valenzuela-Soto, E. M., (2000). Kinetic study of porcine kidney betaine aldehyde dehydrogenase. *Biochem. Biophys. Res. Commun.* 2000; 269: 596-603.

[25] Valenzuela-Soto, E. M., Velasco-García, R., Mújica-Jiménez, C., Gaviria-González, L., Muñoz-Clares, R. A., (2003). Monovalent cations

requirements for the stability of betaine aldehyde dehydrogenase from *Pseudomonas aeruginosa*, porcine kidney and amaranth leaves. *Chem. Biol. Interact.* 143-144, 139-148.

[26] Valenzuela-Soto, E. M., Ayala-Castro, H. G., Muñoz-Clares, R. A., (2005). Effects of monovalent and divalent cations on the thermostability of mammal betaine aldehyde dehydrogenase. In: Maser E, Plapp B, Lindahl R, Weiner H, eds. Enzymology and Molecular Biology of Carbonyl Metabolism. West Lafayette: *Purdue University Press*, pp 104-109.

[27] Ayala-Castro, H. G., Valenzuela-Soto, E. M., (2007). Ionic strength dependency of urea effect on renal betaine aldehyde dehydrogenase stability. In: Weiner H, Maser E, Lindahl R, Plapp B. eds. Enzymology and Molecular Biology of Carbonyl Metabolism. West Lafayette: *Purdue University Press*, pp 77-82.

[28] Neuhofer, W., Beck, F-X., (2006). Survival in hostile environments: strategies of renal medullary cells. *Physiology* 21, 171-180.

[29] Yancey, P. H., (1988). Osmotic effectors in kidney of xeric and mesic rodents: corticomedullary distributions and changes with water availability. *J. Comp. Physiol.* Part B 158, 369-380.

[30] Neuhofer, W., Muller, E., Burger-Kentischer, A., Fraek, M. L., Thurau, K., Beck, F. X., (1998). Pretreatment with hypertonic NaCl protects MDCK cells against high urea concentrations. *Eur. J. Physiol.* 435, 407-414.

[31] Moriyama, T., García-Perez, A., Burg, M. B., (1990). Factors affecting the ratio of different organic osmolytes in renal medullary cells. *Am. J. Physiol. Renal Physiol.* 259, F787-F858.

[32] Guo, X., Zhang, X., Wang, T., Xian, S., Lu, Y., (2016). 3-Bromopyruvate and sodium citrate induce apoptosis in human gastric cancer cell line MGC-803 by inhibiting glycolysis and promoting mitochondria-regulated apoptosis pathway, *Biochem. Biophys. Res. Commun.* 475, 37-43.

[33] Xu, H. M., Tam, N. F. Y., Zan, Q. J., Bai, M., Shin, P. K. S., Vrijmoed, L. L. P., Cheung, S. G., Liao, W. B., (2014). Effects of salinity on anatomical features and physiology of a semi-mangrove plant Myoporum bontioides. *Mar. Pollut. Bull.* 85, 738-746.

[34] Ayala-Castro, H. G., Valenzuela-Soto, E. M., Figueroa-Soto, C. G., Muñoz-Clares, R. A., (2007). Complex conformational changes in kidney betaine aldehyde dehydrogenase revealed by modification with disulfiram. *Arch. Biochem. Biophys.* 468:167-173.

[35] Muñoz-Clares, R. A., González-Segura, L., Mújica-Jiménez, C., Contreras-Díaz, L., (2003). Ligand-induced conformational changes of betaine aldehyde dehydrogenase from *Pseudomonas aeruginosa* and *Amaranthus hypochondriacus* L. leaves affecting the reactivity of the catalytic thiol. *Chem. Biol. Interact.* 143-144, 129-136.

[36] Nordstrom, L. J., Clark, C. A., Andersen, B., Champlin, S. M., Schwinefus, J. J., (2006). Effect of ethylene glycol, urea, and N-methylated glycines on DNA thermal stability: the role of DNA base pair composition and hydration. *Biochemistry* 45, 9604-9614.

[37] Rawling, J. M., ApSimon, M. M., Kirkland, J. B., (1996). Lung poly (ADP-ribose) and NAD^+ concentrations during hypoxia and niacin deficiency in the Fischer-344 rat. *Free Radic. Biol. Med.* 20, 865-871.

[38] Rahman, S., Rehman, M. T., Singh, L. R., Warepam, M., Ahmad, F., Dar, T. A., (2015). Salt potentiates methylamine counteraction system to offset the deleterious effects of urea on protein stability and function. PLoS ONE [Online] 2015; 10(3): e0119597. Available at: http://dx.doi.org/10.1371/ journal.pone.0119597.

[39] Stopa, J. D., S., D. R. Tolan, D. R., (2011). Stabilization of the predominant disease-causing aldolase variant (a149p) with zwitterionic osmolytes. *Biochemistry* 50, 663–671.

[40] Auton, M., Rosgen, J., Sinev, M., Holthauzen, L. M. F., Bolen D. W., (2011). Osmolyte effects on protein stability and solubility: A balancing act between backbone and side-chains. *Biophys. Chem.* 159, 90−99.

[41] Kumar, N., Kishore, N., (2014). Protein stabilization and counteraction of denaturing effect of urea by glycine betaine. *Biophys. Chem.* 189, 16-24.

[42] Kim, Y. C., Kwon, D. Y., Kim, J. H., (2014). Alterations in the metabolomics of sulfur-containing substances in rat kidney by betaine. *Amino Acids* 46(4):963-968.

[43] Liu, Y. L., Pan, Y., Wang, X., Fan, C. Y., Zhu, Q., Li, J. M., Wang, S. J., Kong, L. D., (2014). Betaine reduces serum uric acid levels and improves kidney function in hyperuricemic mice. *Planta Med.* 80, 39-47.

[44] Horio, M., Ito, A., Matsuoka, Y., Moriyama, T., Orita, Y., Takenaka, M., Imai, E., (2001). Apoptosis induced by hypertonicity in Madin Darley canine kidney cell: protective effect of betaine. *Nephrol. Dial. Transplant.* 16, 483-490.

In: Dehydrogenases
Editor: Margaret Collins
ISBN: 978-1-53610-550-6

Chapter 3

DEHYDROGENASES IN GLUCOSE METABOLISM AND THEIR ROLES IN DISEASES

Kamonwan Chamchoy and Usa Boonyuen*
Department of Molecular Tropical Medicine and Genetics,
Faculty of Tropical Medicine, Mahidol University, Bangkok, Thailand

ABSTRACT

Dehydrogenases are enzymes that play an important role in oxidation reactions. They oxidize a specific substrate in which one or more hydride ions are transferred to an electron acceptor molecule. Dehydrogenases are involved in many metabolic pathways which maintain normal function and homeostasis of cells in the body. One example is glucose metabolism, which is crucial for providing energy (ATP) as well as other molecules, such as reduced form of nicotinamide adenine dinucleotide phosphate (NADPH) and ribose sugar. In this chapter, we review the 10 dehydrogenase enzymes found in the main pathways of glucose metabolism; glycolysis, lactic acid fermentation, alcoholic fermentation, the citric acid cycle and pentose phosphate pathway. We also describe the role of these dehydrogenase enzymes in diseases. Glucose metabolism is essential to provide metabolic energy to many cells and tissues. Therefore the impairment of function, activity or an alteration in the level of expression of these 10 dehydrogenase enzymes has an impact on the

* Corresponding author: e-mail: usa.boo@mahidol.ac.th, usa.boo@mahidol.edu.

function of many cells and tissues. Convincing evidence has indicated dehydrogenase enzymes are associated with several diseases, including neurological disorders, many types of cancer and blood diseases. In this chapter, we also discuss how dehydrogenases contribute to the etiology of diseases and their clinical manifestations. Additionally, we explore the applications and advantages of using dehydrogenase enzymes in current medical diagnosis and treatment.

INTRODUCTION

Glucose, a major form of available dietary sugar and key metabolite, is the essential energy source for most cells of the body. The study of glucose metabolism is central to cell proliferation, growth, survival, and most recently, tumor progression. The first step of glucose breakdown is glycolysis, where glucose is converted to pyruvate. The main purpose of glycolysis is the generation of energy in the form of ATP. Glycolysis is the metabolic pathway that serves as the foundation of aerobic and anaerobic cellular respiration. In glycolysis, glucose is broken down into two molecules of pyruvate via a ten-step process catalyzed by 10 glycolytic enzymes (Figure 1). In addition to two molecules of pyruvate, glycolysis also yields two molecules of ATP and two high energy electron carrying molecules, NADH.

An alternative pathway for complete glucose breakdown is the hexose monophosphate shunt or pentose phosphate pathway (PPP), composed of oxidative and non-oxidative phases. Rather than production of energy, in the 3 step oxidative phase, cytosolic glucose 6-phosphate is transformed into ribose sugar, a precursor for the biosynthesis of nucleic acids, and in the process, the reducing power of all cells, NADPH, is produced. The non-oxidative phase of the PPP comprises of five steps and serves to convert the 5-carbon sugar into the 6-carbon sugar (Figure 2).

Typically, two ATPs are directly produced in glycolysis, but much more ATP is formed downstream of glycolysis through the complete oxidation of pyruvate. Under aerobic conditions, following glycolysis, pyruvate is transformed into acetyl CoA by pyruvate dehydrogenase (PDH) enzyme complex (Figure 3).

Thereafter, acetyl CoA is oxidized through the citric acid cycle, producing two molecules of CO_2, one molecule of GTP, one molecule of $FADH_2$ and three molecules of NADH. The citric acid cycle is a series of reaction consisting of eight steps (Figure 4). It is the central metabolic hub of cells and is an important source of precursors for fuel storage and biosynthesis of various molecules, including proteins, carbohydrates and cholesterol. Eventually, $FADH_2$ and NADH produced from the oxidative decarboxylation of glucose are shunted through oxidative phosphorylation.

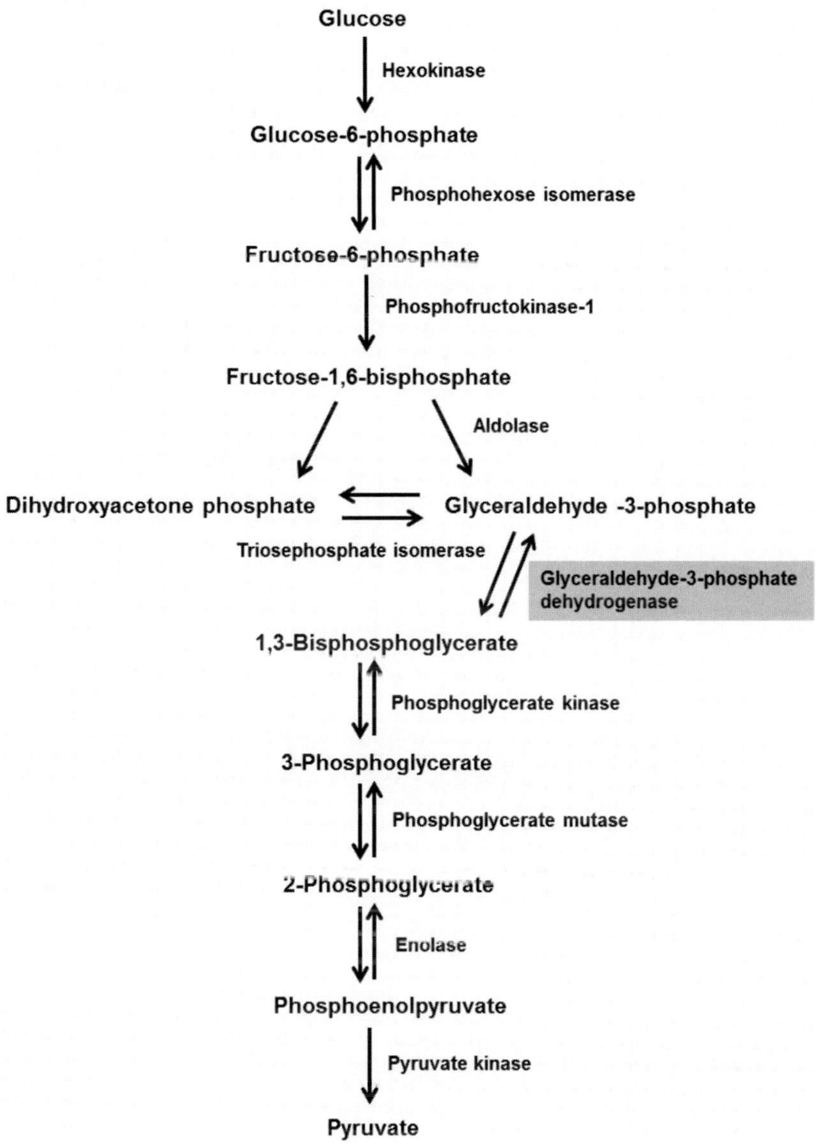

Figure 1. Glycolysis pathway.

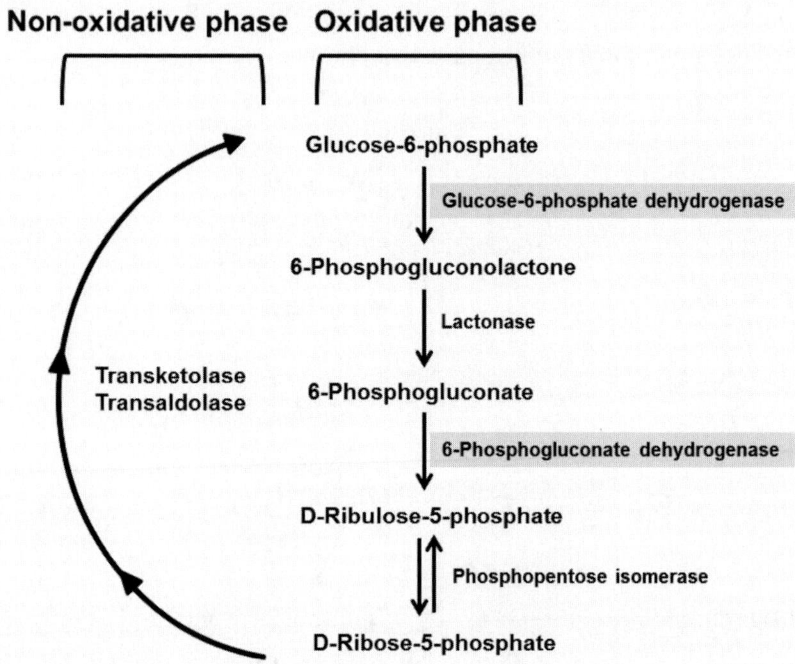

Figure 2. Pentose phosphate pathway.

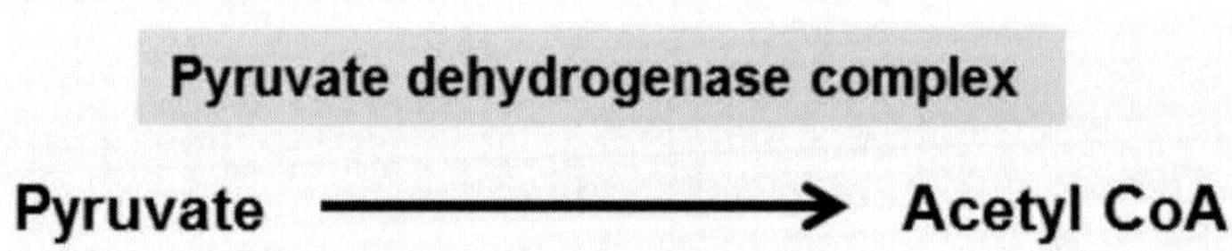

Figure 3. Transformation of pyruvate to acetyl CoA.

Under anaerobic conditions, pyruvate is turned into lactate by the action of lactate dehydrogenase (LDH). Under hypoxic conditions, pyruvate can be changed into ethanol in a two-step process sequentially catalyzed by pyruvate decarboxylase and alcohol dehydrogenase (Figure 5).

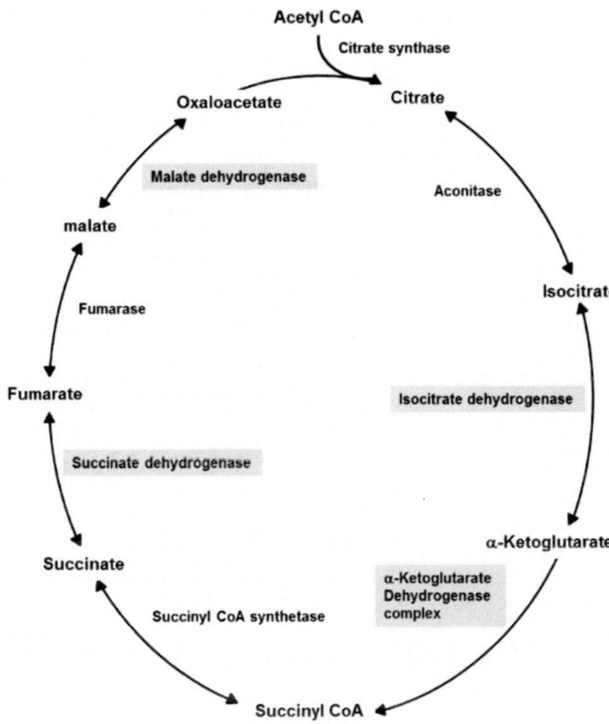

Figure 4. Citric acid cycle.

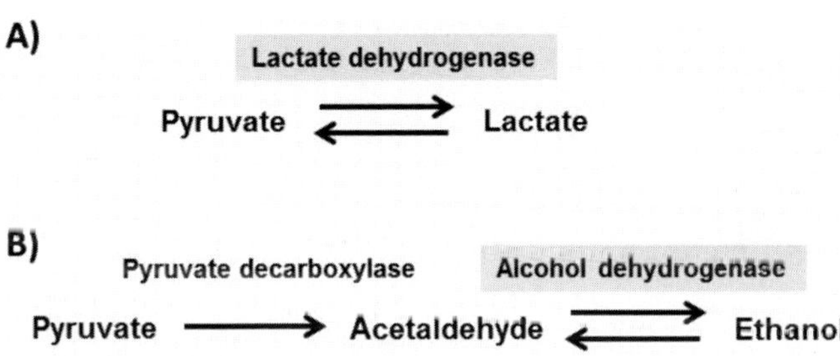

Figure 5. Fate of pyruvate under hypoxia.

Dehydrogenases belong to the group of oxidoreductases (EC 1) that catalyzes the oxidation of a specific substrate in a reduction reaction by which one or more hydrogens are transferred from the substrate, an electron donor, to an electron acceptor molecule or coenzyme such as NAD^+, $NADP^+$, FMN and FAD. There are 10 dehydrogenase enzymes catalyzing the reactions in glucose metabolism:

Glycolysis: glyceraldehyde-3-phosphate dehydrogenase (GAPDH)

Pentose phosphate pathway: glucose-6-phosphate dehydrogenase (G6PD) and 6-phosphogluconate dehydrogenase (6PGD)

Conversion of pyruvate to acetyl coenzyme A (CoA) under aerobic condition: pyruvate dehydrogenase complex (PDHc)

Conversion of pyruvate to lactate or ethanol under anaerobic condition: lactate dehydrogenase (LDH) and alcohol dehydrogenase (ADH)

Citric acid cycle: isocitrate dehydrogenase (IDH), α-ketoglutarate dehydrogenase (KGDH), succinate dehydrogenase (SDH) and malate dehydrogenase.

Becuase cellular metabolism and proliferation are largely dependent on the availability of energy (ATP) produced by glucose metabolism, alteration in glucose metabolism resulting from enzyme defects contributes to various disease conditions. In particular, dehydrogenase enzymes in glucose metabolism have been shown to play potential roles in the etiology of several types of cancer, neurodegenerative disorders and blood diseases. In this chapter, the role of all 10 dehydrogenase enzymes involved in glucose metabolism will be reviewed, including their contribution to the cause of disease as well as their clinical applications.

GLYCERALDEHYDE-3-PHOSPHATE DEHYDROGENASE

Glyceraldehyde-3-phosphate dehydrogenase (GAPDH; EC 1.2.1.12), a key enzyme in the glycolytic pathway, catalyzes the conversion of glyceraldehyde-3-phosphate to 1, 3-bisphosphoglycerate, using NAD^+ as an electron acceptor (Figure 6). GAPDH is a homotetrameric enzyme containing four identical subunits. Although GAPDH is commonly used as an internal control in gene expression and protein studies, it was recently been reported to be altered in various diseases [1, 2]. GAPDH expression is regulated at the transcriptional level and it undergoes post-translational modification. Several transcription factors have been reported to bind to the control region of *GAPDH* promoter to regulate its mRNA expression. Moreover, GAPDH contains specific amino acid residues that are phosphorylation targets for kinases, and phosphorylated-GAPDH is the active form of GAPDH [2].

In addition to its basic function in glucose metabolism, GAPDH has been implicated in diverse cellular processes.

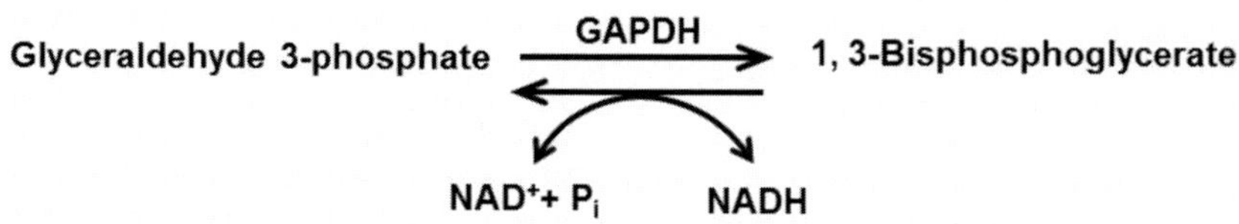

Figure 6. Reaction catalyzed by GAPDH.

Glyceraldehyde-3-Phosphate Dehydrogenase and Cancer

Cancer cells exhibit altered metabolism, which is characterized by a preference for energy production from aerobic glycolysis rather than mitochondrial oxidative phosphorylation. The common features of this altered metabolism are dramatically increased glucose uptake and lactate fermentation even in the presence of an adequate oxygen supply, a process known as the Warburg effect. This alteration has been correlated with tumor progression including the proliferation, migration, metastasis, and viability. However, the mechanism(s) involved in this phenotype are still unclear. Because GAPDH has a role in glycolysis and mitochondrial oxidation by providing substrate to initiate the citric acid cycle or tricaboxylic acid (TCA) cycle, it was proposed that GAPDH might contribute to the Warburg effect [3].

GAPDH expression is deregulated in some cancers and is commonly upregulated in different cancer cells. The overexpression of GAPDH supports cancer cell growth and survival because most cancers preferentially use glycolysis for energy metabolism, suggesting its involvement in Warburg effect. Clinical tissue studies showed that the levels of GAPDH protein were significantly upregulated in lung squamous cell carcinoma tissues. GAPDH knockdown by small or short interfering (si)RNA resulted in the significant reduction of proliferation, migration, and invasion of lung squamous carcinoma cells *in vitro*. A complementary study in a nude mouse cancer xenograft model reported that GAPDH knockdown also significantly inhibited cancer cell proliferation, migration, and invasion *in vivo* [4]. These results suggest that the increased expression of GAPDH enhances glycolytic capacity, thus facilitating tumor progression. The association between increased GAPDH and tumor progression or the loss of GAPDH and reduced cancer cell survival has been demonstrated in several cancers, including hepatocellular carcinoma (HCC), pancreatic cancer, prostate cancer, and colorectal cancer [5-8].

Accordingly, GAPDH is considered a promising biomarker and therapeutic target for cancer. Wang et al. demonstrated that the upregulation pattern of GAPDH was proportional to the malignant stage, because its expression was associated with the level of genes involved in metabolism and cell cycle arrest. Moreover, GAPDH expression was positively correlated with severe solid tumors, suggesting it has predictive power for unfavorable prognoses [9]. Indeed, a correlation between increased GAPDH expression and poor prognosis was revealed in non-small cell lung cancer (NSCLC) [10].

The upregulation of GAPDH leads to enhanced cancer cell metabolism, which is correlated with increased tumor progression and poor patient prognosis in several cancers; however, the molecular mechanisms that underlie how GAPDH mediates cancer cell survival is poorly understood. In a recent study by Colell et al. the role of enhanced GAPDH in cancer survival was established. Changes in mitochondrial outer-membrane permeabilization (MOMP) leads to the release of cytochrome c, which is followed by cell apoptosis induced by caspase activation. However, MOMP can also directly stimulate apoptosis without the need for caspase molecules, termed caspase-independent cell death (CICD). It was reported that the overexpression of GAPDH protected cells from CICD. GAPDH increases glycolysis by providing ATP support for mitochondrial autophagy, resulting in the protection of cells from CICD [11]. Another study showed that GAPDH modulated the activity of cyclin B-cdk1, a protein kinase controls the cell cycle, resulting in increased mitosis and accelerated tumor cell cycle progression [12].

The function of GAPDH was reported in a study using ovarian cancer cell lines to investigate the interaction between GAPDH and colony-stimulating factor-1 (CSF-1), an important cytokine that promotes cell proliferation and differentiation, which is strongly correlated with a poor prognosis of patients with ovarian cancer. GAPDH siRNA reduced both CSF-1 mRNA and protein levels by decreasing the CSF-1 mRNA half-life by 50%. Immunohistochemical analysis of tumor samples from patients demonstrated GAPDH was deregulated and was associated with CSF-1 score [13]. This study demonstrated that GAPDH interacts and controls the stability of mRNA, consequently stabilizing the mRNA of genes involved in tumor progression.

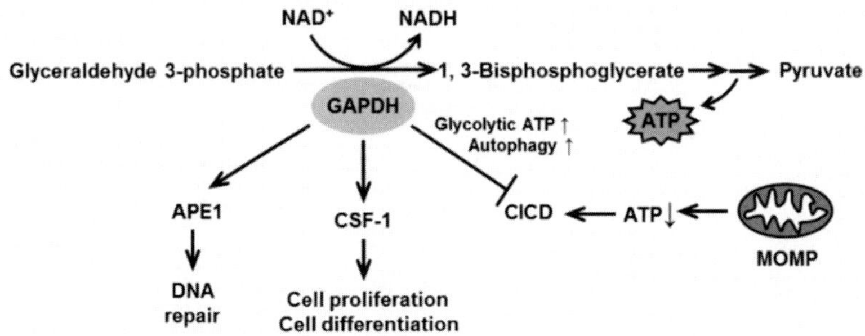

Figure 7. The role of GAPDH in cancer progression. GAPDH inhibits the caspase-independent cell death (CICD) by providing cellular energy and increasing mitochondrial autophagy. GAPDH also activates the function of colony-stimulating factor-1 (CSF-1) and apurinic/apyrimidinic endonuclease-1 (APE1), resulting in promotion of DNA repair and cell survival.

Another function of GAPDH is protection against damaged DNA and DNA degradation. Under normal physiological conditions, GAPDH repairs and/or removes denatured DNA. Demarse et al. demonstrated that GAPDH binds to telomeric DNA directly *in vitro* and may have a biological role in the protection of telomeres against rapid degradation [14]. A study by Azam et al. revealed that GAPDH interacts with apurinic/apyrimidinic endonuclease-1 (APE1), an essential enzyme involved in the repair of abasic sites in damaged DNA. This interaction leads to the reactivation of APE1 endonuclease activity, preventing genomic instability caused by aberrant structural changes [15]. These studies indicate that GAPDH is involved in the induction of DNA repair and prevention of DNA disposal, mechanisms that mediate cancer cell progression as well as make cancer cells resistant to several DNA-targeted chemotherapies.

Taken together, GAPDH can act as both an anti-apoptosis and pro-survival molecule, both of which encourage the progression of several cancers, Figure 7.

Glyceraldehyde-3-Phosphate Dehydrogenase and Neurodegenerative Disease

The major contribution to brain pathogenesis is the excessive apoptosis of neuron cells. A large number of studies have reported the role of GAPDH in apoptosis, suggesting the potential participation of GAPDH in neurodegenerative diseases. Three major neurodegenerative diseases are associated with GAPDH function, including Parkinson's disease (PD), Alzheimer's disease (AD), and Huntington's disease (HD).

PD is a movement disorder characterized by the degeneration of dopaminergic neurons in the substantia nigra. Tatton et al. examined the subcellular localization of GAPDH in nigral neurons from PD patients and found it was present in the cytoplasm of the substantia nigra in the control group. In contrast, GAPDH nuclear accumulation was observed, in addition to increased immunoreactivity of caspase 3 and Bax, in melanized neurons of the PD substantia nigra [16]. This suggested that the accumulation of GAPDH in neural nuclei might mediate apoptotic signaling pathways, resulting in the degeneration of neurons in the substantia nigra.

In a study of rotenone-induced PD rats, immunohistochemistry combined with western blot analysis revealed the overexpression of GAPDH. Overexpression of GAPDH was found in TUNEL-positive dopaminergic cells, indicating upregulation of GAPDH in apoptotic neurons. In addition, an *in vitro* study of rat adrenal pheochromocytoma cells revealed the reduced pathogenesis of rotenone toxicity in GAPDH knockdown cells. These *in vitro* and *in vivo* data also suggest that GAPDH contributes to the pathogenesis of PD [17].

Symptoms of AD include dementia and memory loss. AD is characterized by the deposition of β-amyloid precursor protein (β-APP) in the cerebral cortex and hippocampus. β-APP induces reactive oxygen species (ROS) production, resulting in the destruction of various biomolecules. The recognition of GAPDH by a monoclonal antibody raised against amyloid plaques from the brains of AD patients indicated that GAPDH interacts with β-APP and is also involved in the development of AD [18]. The consequence of interactions between β-APP and GAPDH was demonstrated by Mazzola and Sirover. The study used fibroblasts from an AD patient and revealed that interactions with β-APP might alter the GAPDH structure, resulting in diminished GAPDH glycolytic activity by 27% [19, 20]. Other studies using animal models reported significant decreased GAPDH activity in brain samples from AD mice. GAPDH interacted with β-amyloid and thiorphan, substances involved in β-amyloid catabolism, as well as lipopolysaccharide and interferons-γ, which can influence the progression of AD, resulting in a pronounced reduction of enzyme activity [21]. Taken together, it was concluded that a defect in energy metabolism mediated by the reduction of GAPDH activity in AD might contribute to neuronal pathogenesis.

Huntington's disease (HD) is a neurodegenerative disorder that causes uncontrolled movement, cognitive impairment, and depression. It is characterized by a loss of neurons in the striatum and cortex caused by neurotoxicity of the mutant huntingtin protein (mHtt) [22]. mHtt was reported to bind with GAPDH, suggesting that interactions between mHtt and GAPDH might be involved in the pathology of HD. [23]. Several studies have investigated the influence of the mHtt-GAPDH complex in HD. Mazzola and Sirover reported that the binding of mHtt with GAPDH reduced GAPDH glycolytic activity by 33% [19, 24]. Immunohistochemical analysis of GAPDH revealed that GAPDH was predominantly distributed in the cytoplasm of neurons throughout the brain from wild-type mice. In contrast, GAPDH was increased in specific brain regions in transgenic mice, including the caudate putamen, globus pallidus, neocortex, and hippocampus, as well as accumulating within the neural nucleus. Moreover, the accumulation of GAPDH was associated with a loss of neurons in the caudate putamen and neocortex [25]. Based on these results, nuclear translocation of the mHtt-GAPDH complex causes the reduction of glycolytic metabolism, which might lead to a change in energy status, resulting in cell death.

Other effects of interactions between mHtt-GAPDH were established by Bae et al. They suggested that pathologies caused by nuclear translocation of the mHtt-GAPDH complex were unrelated to decreased GAPDH glycolytic activity or the energy content of cells. GAPDH acts as a carrier to transport mHtt into the cell nucleus; therefore, depletion of GAPDH reduced the nuclear localization of mHtt. It was concluded that neuron damage in HD is caused by cytotoxicity induced by mHtt; however, the exact mechanisms whereby nuclear mHtt elicits cytotoxicity are still unclear [26].

Glyceraldehyde-3-Phosphate Dehydrogenase and Diabetes

Diabetes has numerous severe complications. Recently, GAPDH was reported to play a role in the development of complications associated with diabetes. Methylglyoxal (MG), an α-ketoaldehyde elevated in diabetes, is toxic to cells. The production of MG occurs primarily from intermediates during glycolysis. Therefore, GAPDH is considered to participate in the formation of MG. Treatment of a human red blood cell culture with koningic acid, an inhibitor of GAPDH, caused a 79% reduction in GAPDH activity. Moreover, a 6-fold increase in MG was observed in addition to enhanced pH. The inverse relationship between GAPDH activity and MG production was also reported in *in vivo* studies of type 1 and 2 diabetes [27]. These studies indicated that change in GAPDH activity might lead to the occurrence of diabetes complications.

Other studies have elucidated the role of GAPDH in the development and progression of diabetic retinopathy using a diabetic induced rat model. Rats with poor control, classified based on hyperglycemia and histopathology of the retinal microvasculature, showed decreased retinal GAPDH activity and expression as well as increased ribosylation and nitration. GAPDH activity was decreased in both the cytosol and nuclear fractions, but its protein expression was significantly elevated only in the nuclear fraction [28]. However, the use of ribosylation and nitration inhibitors recovered GAPDH activity [29]. Succination was also reported to inactivate GAPDH activity. Increased succination of GAPDH in muscles from diabetes rats was observed along with decreased GAPDH specific activity [30].

Taken together, these studies provide evidence that during the progression of diabetic retinopathy, GAPDH activity is inhibited. The modification of GAPDH can lead to a loss of its activity and nuclear translocation, contributing to the cell apoptosis cascade. Therefore, the activation of GAPDH activity by inhibiting post-translational modifications might be a potential strategy to prevent the development of diabetes complications (Figure 8).

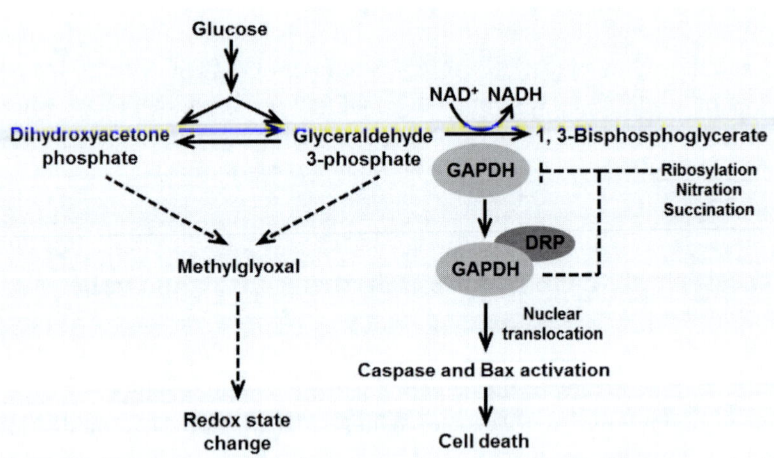

Figure 8. The role of GAPDH in disease development. GAPDH interacts with several disease-related proteins (DRPs). The nuclear accumulation of GAPDH-DRP complex leads to an activation of apoptotic pathway and a reduction in glycolytic activity. When GAPDH is inactivated, dihydroxyacetone phosphate and glyceraldehyde 3-phosphate are converted into methylglyoxal, resulting in enhanced blood pH.

GLUCOSE-6-PHOSPHATE DEHYDROGENASE

Glucose-6-phosphate dehydrogenase (G6PD; EC 1.1.1.49) is a metabolic enzyme that catalyzes the first and rate-limiting step in the PPP, producing NADPH, a reducing power to all cells, and 6-phosphogluconolactone (Figure 9). NADPH plays a crucial role in counterbalancing the oxidative stress in cells. It is necessary for regeneration of reduced glutathione (GSH), which is essential for reduction of hydrogen peroxide and ROS and allowing proteins to be maintained in their reduced states. Additionally, NADPH is required as an electron donor for various enzymatic reactions in other biosynthetic pathways such as fatty biosynthesis.

Glucose-6-Phosphate Dehydrogenase Deficiency

G6PD deficiency is a hereditary genetic defect and is one of the most prevalent polymorphisms and enzymopathies in humans, affecting approximately 400 million people worldwide. It was first discovered in 1926 during malaria treatment in which hemolytic anemia was frequently observed in individuals treated with 6-methoxy-8-aminoquinoline (primaquine) drugs [31]. The gene encoding G6PD enzyme is located at the q28 locus on the X-chromosome. Therefore, the clinical features are more prominent in males. While homozygous females are relative rare, heterozygous females exhibit an array of enzyme activities because of partial inactivation of one X-chromosome. G6PD deficiency is caused by mutations in the *G6PD* gene, leading to G6PD variants with varied enzyme activities. Clinical manifestations of G6PD deficiency can range from asymptomatic, favism, hemolytic anemia, chronic non-spherocytic hemolytic anemia (CNSHA), and spontaneous abortions to neonatal hyperbilirubinemia resulting in neonatal kernicterus, which could lead to death [32-34]. However, the most common clinical symptoms are neonatal jaundice and acute hemolytic anemia, which are usually triggered by oxidative agents.

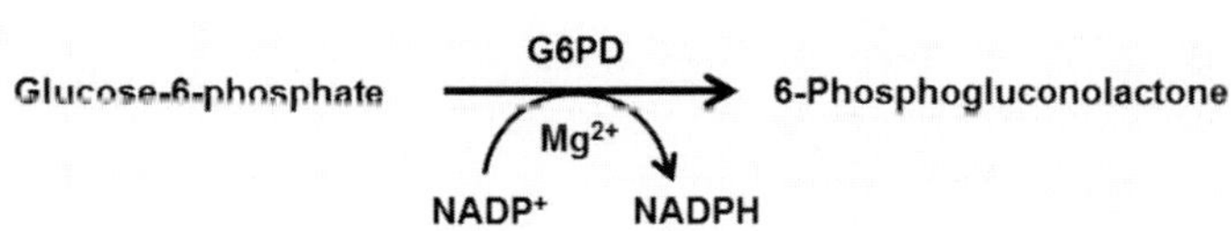

Figure 9. Reaction catalyzed by G6PD.

WHO working groups have classified G6PD variants, based on enzyme activity, into five groups with thresholds of 10–60% as the defined level of G6PD activity. Class I variants cause severe deficiency, and have less than 10% activity with CNSHA. Class II variants also cause severe deficiency and have less than 10% activity remaining with intermittent hemolysis. Class III variants cause mild deficiencies with 10–60% enzyme activity resulting in hemolysis only when exposed to triggers (oxidative stressors such as certain drugs, ingestion of fava beans and infections). Class IV variants are non-deficient variants with 60–100% enzyme activity and no clinical features. Class V variants are those with increased enzyme activity and no clinical manifestation. Fortunately, class I variants are relatively rare and most of the common variants are asymptomatic.

G6PD deficiency is found widespread in malaria endemic areas [35-40]. The geographical correlation of the distribution of G6PD deficiency with the endemicity of malaria suggests that it might arise through the natural selection of malaria [41]. There are reports of the protective effect of G6PD deficiency against *Plasmodium falcuparum* and *P. vivax* from both epidemiological correlations and *in vitro* studies [42, 43]. Currently, more than 180 G6PD deficiencies were identified at the DNA level and over 400 variants were described based on biochemical variants [44-46]. Most of the variants identified are single mutations while multiple mutations and deletion mutations are found with lower frequency. G6PD deficiency was characterized by measuring several biochemical properties, including enzyme activity, kinetic properties, electrophoretic mobility, thermostability and chromatographic behavior. The reduced enzyme activity of G6PD deficiency can be caused by a decrease in catalytic activity or a reduced number of active enzyme molecules. Changes in catalytic activity mostly results from mutations in the *G6PD* gene and depends on the mutation position. Reduction of active G6PD molecules could arise from protein instability, defects in protein folding, failure to form stable dimers and trouble with mRNA splicing. To understand the molecular mechanisms underlying the reduced G6PD activity of G6PD deficiency, several G6PD variants have been characterized in terms of their biochemical properties, stability and structure [45, 47-50].

Because NADPH is required to counterbalance the oxidative stress in which it is necessary for reduction of hydrogen peroxide and ROS and allowing proteins to be maintained in their reduced states, deficiency of G6PD leads to an accumulation of ROS in cells (Figure 10).

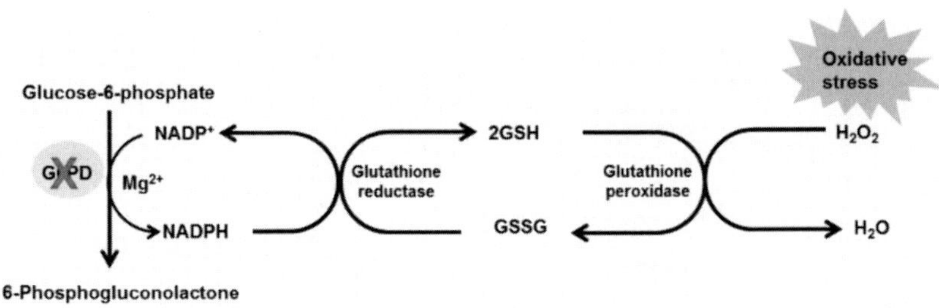

Figure 10. G6PD deficiency. G6PD produces NADPH, a reducing power to all cells. Because NADPH is required to counterbalance the oxidative stress via the regeneration of GSH, deficiency in the G6PD enzyme results in accumulation of oxidative stress in cells. G6PD deficient cells are susceptible to oxidative agents such as drugs, causing hemolytic anemia. G6PD deficiency also promotes viral infection.

G6PD deficiency usually presents as drug-induced or infection-induced hemolytic anemia, neonatal jaundice and/or CNSHA. Drug-induced hemolytic anemia in G6PD deficiency individuals varies and depends on factors such as pharmacokinetics, underlying clinical conditions and type of medication. Furthermore, hemolysis in G6PD deficiency is a self-limiting process, and therefore, it does not necessarily result in clinical anemia. Infections by hepatitis viruses A and B, pneumonia and typhoid fever also cause hemolytic anemia in G6PD deficient individuals with varying degrees of severity depending on administered drugs, age and liver function. Additionally, several reports have described the association of G6PD deficiency with other clinical disorders such as neonatal hyperbilirubinemia, heart dysfunction, diabetics, neurodegenerative conditions and cataracts. Though G6PD deficiency and sickle cell anemia are genetically unrelated, an association between them has been reported [51]. When G6PD deficiency and sickle cell anemia co-exists in one individual, it is believed to provide a selective advantage against malaria infection [52]. G6PD deficiency was proposed as a risk factor for kernicterus as it was observed that the incidence of severe neonatal hyperbilirubinemia is several folds greater in the G6PD deficient compared with the G6PD normal population [53]. An interaction between G6PD deficiency and mutation in the UTG promoter associated with Gilbert's syndrome was observed, leading to hyperbilirubinemia in neonates. Neither G6PD deficiency nor UTG promoter mutation alone can cause hyperbilirubinemia. It was speculated that increased hemolysis in G6PD deficient neonates in combination with uneven bilirubin metabolism caused an accumulation of bilirubin resulting in hyperbilirubinemia.

G6PD deficiency was reported to promote the infection of enterovirus 71 (EV71). G6PD deficient fibroblast cells are more susceptible to EV71 infection when compared with normal cells. The efficiency of virus replication in G6PD deficient cells is also greater than that in wild-type cells, indicating that the redox status plays an important role in the susceptibility of viral infection [54]. In G6PD deficient cells, ROS production is increased which might regulate the signaling pathway, through mitogen-activated protein kinase (MAPK), causing elevated replication of the virus [55, 56]. Additionally, G6PD deficiency also enhanced the infection of human coronavirus 229E (HCoV-229E) [57]. An increased rate of infection was associated with increased ROS production in host cells. The presence of antioxidants diminished the oxidative stress in cells, reducing the susceptibility to viral infection.

Additionally, a higher infection of dengue virus serotype 2 to monocyte in G6PD deficient patients was reported [58]. This suggested that the environmental state of the host including the level of oxidative stress plays a critical role in pathogen infectivity. As a consequence, the mechanism underlying the susceptibility of G6PD deficiency to viral infection was investigated [59]. Expression of antiviral genes such as tumor necrosis factor α (TNF-α) and GTPase myxovirus resistance 1 (MX1) was decreased in G6PD knockdown A549 cells upon infection by EV71 and HCoV-229E. Furthermore, it was found that nuclear factor-κB (NF-κB) binding activity toward TNF-α and MX1 was diminished in G6PD deficient cells upon viral infection, suggesting that activation of NF-κB is essential for induction of antiviral gene expression. Inhibition of TNF-α has resulted in enhancement of viral replication and increases in viral gene expression in A549 cells, indicating that TNF-α plays a major role in the antiviral response. Therefore, a reduction of TNF-α might contribute to increasing viral replication in these cells.

The redox status in G6PD knockdown cells was monitored by measuring the expression of NmrA-like family domain containing protein 1 or NMRAL1 (HSCARG), a cellular redox sensor of the NADPH/NADP ratio and a regulator of NF-κB signaling. Expression of HSCARG was increased in G6PD deficient cells, affecting NF-κB signaling and the antiviral response. The results from this study indicated that G6PD deficiency impaired the innate immune response to viral infection through the HSCARG- NF-κB pathway, causing higher susceptibility to viral infection.

Glucose-6-Phosphate Dehydrogenase and Cancer

G6PD activity or its expression is highly increased in many types of cancer cells, including those of the gastrointestinal tract, colon, uterus and lung as well as B-cells. Therefore, it was proposed that G6PD is important for proliferation and growth of cancer cells. In supporting this notion, mechanisms by which G6PD contributed to cancer have been extensively investigated.

In HCC, binding of phosphatase and tensin homologue (PTEN) protein to G6PD prevented the formation of the active dimer [60]. PTEN also blocked G6PD pre-mRNA splicing by interacting with heterogeneous nuclear ribonucleoprotein K (hnRNPK). This resulted in reduction of G6PD activity, affecting glucose consumption and biosynthesis. However, T cell leukemia 1 (Tcl1) reversed the action of PTEN by activating G6PD activity, increasing pre-mRNA splicing of G6PD and promoting G6PD expression via a hnRNPK mechanism. Knockdown of G6PD inhibited hepatocarcinogenesis. G6PD activity was inhibited in the presence of B-cell lymphoma 2 (Bcl-2) associated athanogene 3 or BAG3, resulting in cell proliferation suppression [61]. It was also reported that both short term and long term G6PD activity were stimulated by growth factors such as epidermal growth factor. Furthermore, it was found that G6PD activity is crucial for controlling intracellular redox during cell growth in which NADPH is required for the antioxidative response by controlling the concentration of GSH in the cell. Additionally, the clinical relevance of G6PD in HCC treatment was determined in patients who received sorafenib after liver cancer surgery. The low expression levels of G6PD and Tcl1 predicted that these patients would have a better survival rate, suggesting the importance of these proteins in the pathology of HCC [60].

In lung cancer, p73, a protein required for cell proliferation, stimulated the expression of G6PD in human lung cancer H1299 and mouse tumor cells [62]. Silencing p73 resulted in reduction of G6PD expression and cell proliferation in human osteosarcoma U2OS cells, mouse embryonic fibroblast, isogenic human cancer HCT116 cells and human lung cancer H1299 cells. Suppression of p73 in human lung cancer H1299 cells decreased NADPH and DNA synthesis. In the absence of p73, the expression level of G6PD decreased in cells experiencing DNA damage. Additionally, knock down of G6PD resulted in reduced NADPH level, DNA synthesis and cell proliferation [63].

In cervical cancer patients, the expression level of G6PD was increased and also associated with the development of human papillomavirus (HPV) infection. Down regulation of G6PD reduced cell proliferation while inducing more apoptosis in HPV18-Hela cells [64]. Additionally, the relationship between antioxidant enzymes, including G6PD and the oxidative stress status in erythrocytes of cervical cancer patients was determined. The activity of G6PD and other antioxidant enzymes was decreased in the erythrocytes of cervical cancer patients, suggesting oxidative damage occurs in the erythrocytes of cervical cancer patients [65].

In breast cancer, *in vitro* and *in vivo* experiments suggested that oxidative stress might be involved in initiation and/or progression of breast neoplasma via estrogen-mediated oxidative DNA damage. The activity of G6PD and other antioxidant enzymes was significantly increased when MCF-7 cells were treated with 17β-estradiol [66]. G6PD expression was significantly increased in primary breast carcinoma and levels of G6PD expression were significantly associated with progression-free survival. Patients with low G6PD expression were more likely to live longer with no recurrence or metastasis. Therefore, it was suggested that G6PD could serve as a prognostic marker for primary breast carcinoma and expression of G6PD could be an indicator of recurrent metastasis [67]. In supporting this notion, high expression of G6PD indicates a poor survival rate [68]. G6DP was found to be closely correlated with the molecular subtypes, grade and stage of breast cancer. It was also suggested that crosstalk might exist among G6PD and the M2 isoform of pyruvate kinase, glutamine-fructose-6-phosphate transaminase 1 and p53 proteins in breast cancer.

A proteomic approach was employed to investigate the mechanism of exosome-regulated epithelial ovarian cancer. Enzymes in PPP, including G6PD, were markedly upregulated in the late-stage ovarian cancer cells, HO8910PM and OVCA429 [69]. Additionally, G6PD expression and activity were accelerated in cisplatin-resistant ovarian cancer cells when compared with sensitive cells. A combination of cisplatin and G6PD inhibitors showed a positive effect on resistant cells, suggesting G6PD may be a potential therapeutic target in cisplatin-resistant ovarian cancer cells [70].

The molecular mechanism of G6PD in esophageal squamous cell carcinoma (ESCC) was also elucidated. The mRNA and expression level of G6PD was elevated in ESCC cells when compared with normal cells [71]. Expression of G6PD was also correlated with histological grade, lymph node metastasis and the stage of ESCC, suggesting G6PD was an independent prognostic factor for ESCC. Depletion of G6PD *in vitro* and *in vivo* resulted in growth suppression, which was correlated with decreased tumor weight as well as the Ki-67 proliferation index. Inhibition of G6PD also caused the cell cycle to arrest at the G0/G1 phase together with the reduction of cyclin D1 and cyclin-dependent kinase 4 (CDK4) protein levels. Additionally, G6PD repression triggered cell apoptosis where the expression of Bcl-2 protein decreased while the activity of caspase-3 and Bax protein expression increased. Interestingly, the phospho- Signal transducer and activator of transcription 3 (STAT3) protein level decreased significantly without a decrease in the total STAT3. These data suggested that G6PD plays an important role in the regulation of ESCC development and progression, possibly, through the STAT3 signaling pathway.

Knockdown of G6PD in A375 human melanoma cells reduced cell proliferation and increased apoptosis. Additionally, suppression of G6PD caused the tumor cells to become more susceptible to oxidative stress, reducing the NADPH level by 20% while the wild-type cells only showed 8.7% reduction in NADPH [72]. The regulation mechanism of G6PD and pathological changes were further investigated in a human melanoma xenograft mouse model [73]. A375 cells transfected with wild-type and mutant G6PDs (deletion and overexpression mutants) were injected in nude mice and expression of G6PD and other proteins related to the cell cycle and apoptosis was examined. G6PD expression and its activity were markedly higher in nude mice tumor tissues when compared with the normal tissues. It was revealed that inhibition of G6PD resulted in a significant decrease of cell cycle-related proteins such as cyclin D1, cyclin E, p53 and S100A4. Bcl-2 and Bcl-x/l, two apoptosis inhibitory factors, were found to increase while the apoptosis factor Fas decreased in knockdown G6PD mice. Interestingly, in agreement to what was observed in ESCC, expression of phospho-STAT3, STAT3, phospho-STAT5 and STAT5 in tumor tissues was reduced in the G6PD suppression group. Furthermore, it was confirmed that suppression of G6PD induced cell arrest at the G1/S border and significantly down-regulated cyclin D1 and CDK4 [74]. However, no difference in ROS was detected in G6PD-suppressed A375 cells when compared with normal cells. Inhibition of G6PD caused a significant increase in STAT3 levels but a reduction in phospho-STAT 3 as a result of increased proto-oncogene tyrosine-protein kinase Src or c-Src and decreased Src-homology 2 domain-containing phosphatase 2 levels. Cyclin D1 functions by promoting the transition of the G1/S phase and enhancing cell division and cell growth. Therefore, overexpression of this protein is considered to be essential factor for tumor occurrence and progression. In melanoma cells, expression of cyclin D1 was elevated, accelerating cell proliferation and facilitating tumor malignancy. The data suggested that G6PD expression in A375 cells plays a vital role in antioxidative responses, tumor growth and proliferation. G6PD can regulate the cell cycle by affecting the expression of cell cycle-related proteins such as cyclin D1, CDK4, cyclin E, p53 and S100A4 and therefore, can control melanoma cell growth.

In lymphoma cancers, it was suggested that oxidative stress resistance and treatment failure could depend on the activity of G6PD. Research in WEHI7.2 cells demonstrated that overexpression of G6PD led to an increase loss of cell viability and also activity of caspase 3, an apoptosis-related protein, when treated with the synthetic glucocorticoid, dexamethasone. This indicated that chemotherapeutic treatment of lymphoma induced the expression of G6PD, leading to reduction of ATP and increased ROS, which in turn, triggered cell apoptosis [75].

The metabolic profile of human clear cell-renal cell carcinoma (ccRCC) was also studied. It was revealed that the activity of G6PD was elevated in association with metabolic alterations of these cancer cells. Inhibition of G6PD resulted in an extreme decrease of cancer cell proliferation, a reduction of NADPH level and an increased production of ROS [76].

G6PD was also studied in the human colon cancer cell line HT29. Enhanced G6PD activity was observed during the G1 and S phases [77]. Another study revealed that the tumor suppressor p53 directly binds to G6PD and blocks the formation of the active dimer, causing a total loss of enzyme activity. As a consequence, the production of ribose-5-phosphate and NADPH was decreased. Cells lacking the p53 protein showed a significant increase of G6PD dimer while the G6PD monomer was decreased [78]. In multi-drug resistant (MDR) cancer cells, increased activity of G6PD was observed in the doxorubicin-resistant human colon cancer cell line (HT29-DX) together with an increased intracellular level of GSH. This study suggested that redox metabolism plays an important role in MDR, possibly through activation of G6PD activity, and also suggested that chemotherapy outcomes could be improved by blocking upregulation of G6PD [79].

G6PD activity was also found to be elevated in prostate cancer [80]. It was demonstrated that G6PD activation occurred through androgen receptor (AR) signaling in prostate cancer. AR regulates mechanistic target of rapamycin (mTOR) signaling, which then activates sterol-regulatory element binding protein 1 (SREBP1) through phosphorylation of its downstream target p70S6K. SREBP1 regulates the expression of G6PD in HEK293 cells [81]. Importantly, it was also shown that the expression level of G6PD is associated with the progression of cancer *in vivo*. Therefore, it was suggested that the AR may possibly regulate G6PD via an mTOR-mediated mechanism in prostate cancer [82].

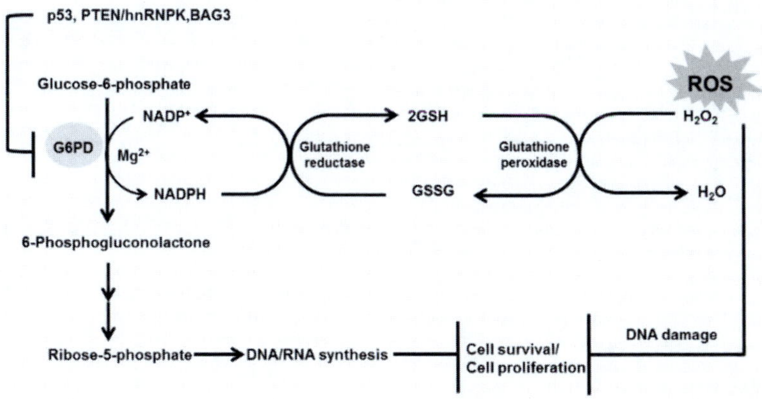

Figure 11. The role of G6PD in tumorigenesis. G6PD expression is regulated by several transcription factors such as p53, PTEN, hnRNPK and BAG3. Because G6PD is essential for cell growth and proliferation as ribose-5-phosphate is a precursor of DNA and RNA biosynthesis, inhibition of G6PD by transcription factors leads to reduced DNA/RNA synthesis, causing cell death. Additionally, inhibition of G6PD enzyme also decreases NADPH production, causing oxidative stress in cells which results in DNA damage and cell death.

Taken together, G6PD plays a crucial role in cell growth, proliferation and survival by producing a precursor for the synthesis of DNA and RNA. Additionally, G6PD generates NAPDH, a reducing power to cells which is required to maintain the redox status. The role of G6PD in tumorigenesis is demonstrated in Figure 11. G6PD is overexpressed in several types of cancer, indicating its contribution in development of cancer.

The role of G6PD in macromolecule biosynthesis and maintenance of cellular redox homeostasis in proliferating cells was revealed. It was found that G6PD is dynamically glycosylated in response to hypoxia [83]. Inhibition of G6PD glycosylation led to the reduction of cell proliferation *in vitro* and decreased *in vivo* cell growth, suggesting that glycosylation is essential for G6PD activity. O-GlcNAcylation of G6PD enhanced its activity, increasing glucose flux through PPP, producing precursors for nucleic acid and lipid biosynthesis and providing NADPH for cell survival under oxidative stress. Interestingly, it was also found that the glycosylation level of G6PD is increased in human lung cancer, indicating that G6PD glycosylation plays an essential role in cell survival and proliferation in cancer cells.

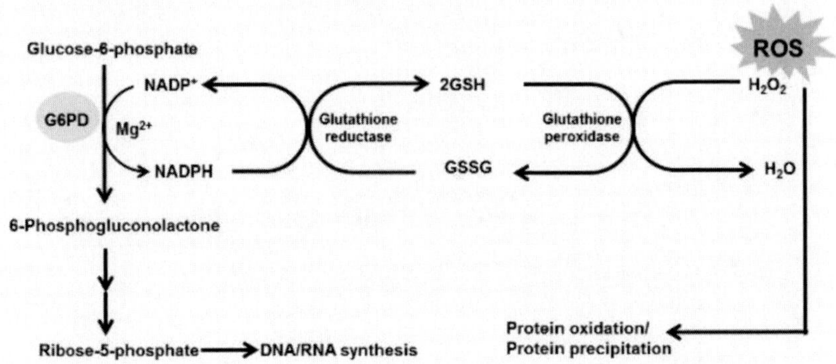

Figure 12. The role of G6PD in cataract. Because NADPH is required to regenerate GSH, maintaining the reduced state of proteins deficiency of G6PD causes an accumulation of oxidative stress, leading to protein oxidation and precipitation in the eye lens.

Regulation of G6PD activity by acetylation has been reported. Acetylation of G6PD at Lys403 caused a complete loss of activity because of its inability to form active enzyme dimers [84]. Additionally, it was revealed that acetylation of G6PD activity occurred via ELP3 histone family acetyltransferase while sirtulin 2 (SIRT2) mediated deacetylation of G6PD, promoting G6PD activity and production of NADPH to combat oxidative stress in mouse erythrocytes.

Glucose-6-Phosphate Dehydrogenase and Cataract

Osmotic and oxidative stresses are thought to play important roles in cataract formation. Reducing power such as that provided by NADPH is required to maintain the reduced state of proteins in the eye lens through the action of GSH. A reduction of the NADPH level would increase the susceptibility of the lens to oxidative stress which subsequently leads to the oxidation of proteins in the lens and precipitation of proteins with a loss in lens transparency [85]. Owing to its ability to produce NADPH, a reducing power, G6PD might play a role in cataracts. It was found that G6PD deficient individuals are at higher risk for developing cataracts than healthy people [86, 87]. Additionally, the levels of G6PD and GSH in senile cataract patients were found to be significantly lower than those in healthy individuals [88]. A study of mRNA expression by real-time polymerase chain reaction (PCR) demonstrated that *G6PD* expression was markedly increased after supplementation of antioxidants [89]. Supplement intake increased the expression of G6PD resulting in an elevated level of NAPDH and an increase in GSH leading to the reduction of oxidative stress in the lens. Altogether, it was suggested that G6PD plays an important role in the development of cataracts. Contribution of G6PD to cataract is illustrated in Figure 12.

Glucose-6-Phosphate Dehydrogenase and Neurodegenerative Disease

It was believed that oxidative stress plays an important role in the pathophysiology of neurodegenerative disorders such as PD and AD, and GSH is essential for the reduction of oxidative stress in neural cells, through the reducing power of NADPH [90, 91]. Oxidative damage to lipids, proteins and DNA was observed in the brains and leukocytes of PD patients [92, 93]. G6PD is a major enzyme providing reducing equivalents to maintain GSH and other sulfhydryl proteins in neurons. It was suggested that G6PD might play a role in neurons and thus, several attempts have been made to understand the function of this enzyme in neurodegenerative disorders and aging. A significant decrease in G6PD activity was described in the erythrocytes of PD patients [94]. Additionally, the levels of G6PD and 6PGD in the putamen and cerebellum of early-stage PD were determined by enzyme-linked immunosorbent assay and a reduction of both enzymes was observed, suggesting a failure in the maintenance of oxidative state in the neurons. Furthermore, the protective effect of G6PD against nitric oxide-mediated apoptosis was reported in pheochromocytoma (PC12) cells [95]. When enhanced G6PD activity was observed, resistance to nitric oxide-mediated apoptosis was observed in PC12 cells.

Other studies also demonstrated the role of G6PD activity in neuroprotection. Transgenic mice overexpressing G6PD in the nigrostriatal system were more resistant to a neurotoxin when compared with the control group, indicating the association between G6PD and oxidative defense in PD [96]. Another investigation in the brain of aged mice also pointed to the protective effect of G6PD against oxidative stress. Oxidative damage was elevated in the brain region containing neuropathological damage in aged G6PD-deficient mice, especially heterozygous and homozygous deficient individuals. Interestingly, it was found that the loss of one copy of the *G6PD* gene could contribute to a higher risk of neurodegeneration [97]. These results confirm the functional role of G6PD in protecting against oxidative damage and also suggest the potential role of G6PD in preventing neurodegeneration.

Moreover, the role of G6PD in AD was also studied. G6PD and sulfhydryl content were found to be elevated in AD samples, indicating a reductive compensation against oxidative stress in AD [98]. Additionally, increased activity of G6PD and 6PGD was also observed in the inferior temporal cortex of Alzheimer patients in another study [99]. It was suggested that upregulation of both enzymes in PPP is occurred in response to increased prooxidant activity. Taken together, the data demonstrated that G6PD plays a role in maintenance and progression of neurodegeneration associated with AD by decreasing oxidative stress in neurons.

Glucose-6-Phosphate Dehydrogenase, Obesity and Diabetes

Because NADPH is essential for regulation of the redox status in cells and is required for the biosynthesis of fatty acids and cholesterol, it was postulated that NADPH generating enzymes such as G6PD might play a part in lipid metabolism disorders such as obesity and diabetes. Significant increased evidence of the association between the incidence of diabetes and G6PD deficiency was found, suggesting an increased risk of diabetes in G6PD deficient individuals [100, 101]. As a consequence, the importance of G6PD in fatty acid synthesis has been studied. Both the mRNA and protein levels of G6PD were increased in the adipose tissues of obese mice compared with lean mice [102]. Other experiments were performed to understand the role of G6PD in lipogenesis and adipogenesis. Overexpression of G6PD in 3T3-L1 cells promoted lipogenic and adipogenic gene expression. Additionally, G6PD overexpression was shown to stimulate the release of free fatty acid (FFA) and triglyceride (TG) while the cholesterol level did not change. Inhibition of G6PD in 3T3-L1 cells resulted in attenuation of adipocyte differentiation with a small accumulation of lipid droplet and reduction of FFA and TG levels. The effects of G6PD suppression on adipogenic gene expression in 3T3-L1 cells were also elucidated. It was found that the mRNA levels of lipogenic genes robustly declined. These results suggested G6PD activity is necessary for both lipogenesis and adipogenesis in adipose tissues.

Overexpression of G6PD was also shown to alter the expression of adipocytokines and induce insulin resistance in obese animals. The association between G6PD and insulin resistance was investigated in diet-induced obesity mice and it was shown that these mice developed insulin resistance with elevated activity of G6PD. Moreover, the body weight, serum TG and FA of the obese mice were also increased [103]. Therefore, it suggested that high G6PD activity might regulate the levels of pentose and pentosidine, two compounds related to the development of insulin resistance in obesity. *In vivo* inhibition of G6PD in mice relieved obesity-induced insulin resistance [104]. G6PD-deficient mice exhibited alleviated insulin resistance and decreased adipose tissue inflammation i.e., reduction in the expression of inflammatory cytokines and down regulation of proinflammatory macrophages. Additionally, expression of NADPH oxidase was decreased, reducing the production of ROS, while antioxidant genes were upregulated in the adipose tissue of G6PD deficient mice. These studies suggested that G6PD plays a role in the mediation of obesity induced lipid metabolic disorders and insulin resistance in mice models, possibly through regulation of the inflammatory response.

The role of G6PD in obesity was further investigated in human subjects [105]. Concentration of G6PD in serum was markedly increased with increasing body mass index (BMI) in obese subjects. Data analysis also showed a positive correlation of G6PD with BMI and malonyldialdehyde (MDA), suggesting that elevated activity of G6PD could induce the onset of metabolic disorders in obesity, possibly through the induction of oxidative stress and inflammatory signals. Additionally, the level of G6PD in adipose tissue could be a good indicator of obesity. Therefore, G6PD might serve as a ground-breaking therapeutic target for metabolic disorders such as obesity.

In diabetic mellitus (DM), it was recognized that G6PD deficient individuals possess a higher susceptibility to oxidative stress which is one of the risk factors for diabetes. The association of G6PD activity and diabetes was assessed in several population groups, including in Taiwan, and a significant difference of G6PD activity was noticed between diabetic and healthy Taiwan subjects [100]. G6PD activity in diabetic patients was remarkably lower than that in healthy individuals. As a consequence, it was proposed that low G6PD activity is a risk factor for pathogenesis of diabetes in the population of Taiwan.

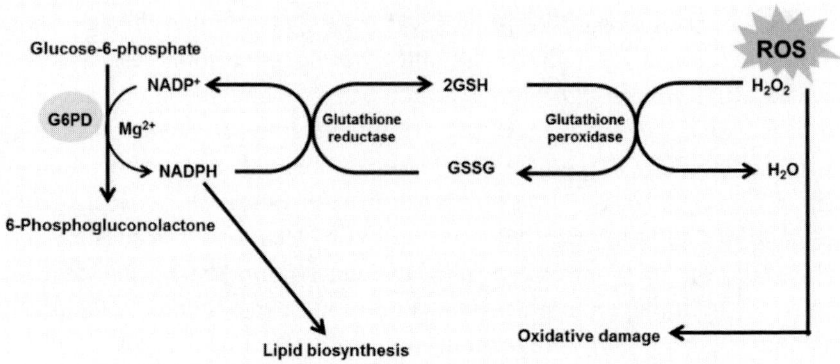

Figure 13. The role of G6PD in obesity. In addition to providing reducing power to all cells, NADPH is required in the lipid biosynthesis. Elevated G6PD activity causes insulin resistance in obesity.

Decreased expression of G6PD was also observed in the induced-diabetic rat [106]. In this study, the diabetic rat was induced by streptozotocin and G6PD activity was determined. Additionally, measurement of oxidative stress markers and protein kinase A (PKA) activity were also performed in the kidney of the diabetic rat. G6PD activity and NADPH and GSH levels were significantly decreased in diabetic animals when compared with the control group. However, lipid peroxidation was found to increase, indicating that diabetes produces more oxidative stress in the kidney cortex. Furthermore, PKA activity was markedly elevated leading to phosphorylation of G6PD, and inhibiting G6PD activity. The expression level of G6PD was decreased in the diabetic rat. The data suggested that both G6PD phosphorylation and reduction of G6PD expression contributed to decreased G6PD activity in diabetic animals, and caused the oxidative stress in the kidney. Several studies have also reported the decline of G6PD activity in different organs of diabetic animals [107, 108]. Moreover, decreased level of G6PD has been reported in diabetic patients [109-111]. Taken together, it was proposed that G6PD plays a major role in pathogenesis of diabetic complications and could possibly be used as a biomarker of diabetic mellitus. The role of G6PD in obesity is shown in Figure 13.

Glucose-6-Phosphate Dehydrogenase and Vascular Disease

Owing to the essential role of G6PD in producing the reducing equivalent, NAPDH, its function in vascular diseases was also examined. Exposure of G6PD deficient endothelial cells to an oxidizing agent resulted in the accumulation of ROS and reduction of GSH, which may lead to cell growth retardation [112]. G6PD was overexpressed in endothelial cells and the association of its activity with oxidative stress and the bioavailability of nitric oxide were determined. G6PD provided a protective effect against oxidative stress in vascular endothelial cells. Overexpression of G6PD was found to decrease ROS accumulation and maintain the level of intracellular GSH. Additionally, G6PD overexpression also activated the activity of the endothelial isoform of nitric oxide synthase (eNOS), leading to nitric oxide formation [113]. This suggested that vascular endothelial cells induced expression of G6PD to protect against oxidative stress and preventing endothelial malfunction. The molecular mechanism underlying the regulation of G6PD in endothelial cells was later investigated [114]. Vascular endothelial growth factor (VEGF) treatment raised G6PD activity and its translocation at the plasma membrane in a tyrosine kinase-dependent manner. After further study, it was discovered that Src is responsible for G6PD tyrosine phosphorylation in response to VEGF stimulation in endothelial cells. This indicated that Src-mediated tyrosine phosphorylation of G6PD is essential for G6PD activity and its regulation could be a potential target for treating tumor angiogenesis.

6-Phosphogluconate Dehydrogenase

6-Phosphogluconate dehydrogenase (6PGD; EC 1.1.1.44) is the third enzyme and second dehydrogenase in the PPP, catalyzing the conversion of 6-phosphogluconate to ribulose 5-phosphate and producing NADPH in the presence of $NADP^+$, Figure 14.

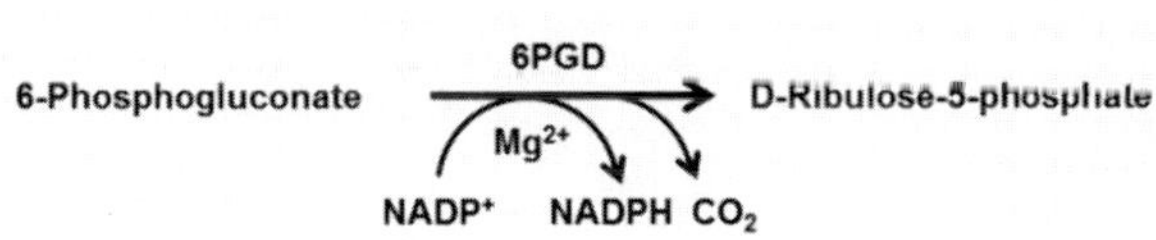

Figure 14. Reaction catalyzed by 6PGD.

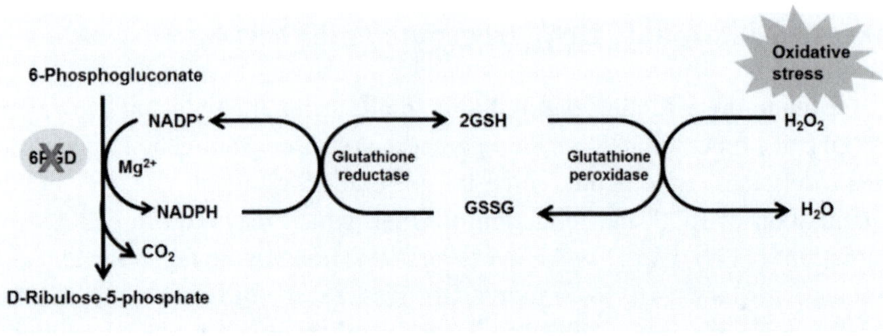

Figure 15. 6PGD deficiency. 6PGD produces NADPH, a reducing power to all cells. Because NADPH is essential to decrease the oxidative stress by regenerating GSH, deficiency in the 6PGD enzyme results in accumulation of oxidative stress in cells.

6-Phosphogluconate Dehydrogenase Deficiency

While G6PD deficiency is the most common enzymopathy in human, 6PGD deficiency is extremely rare (less than 1%). The deficiency is autosomal dominant inheritance that can result in a lack of 6PGD protein synthesis or highly unstable protein or protein that exhibits reduced activity (Figure 15). Several genetic variants of 6PGD have been discovered, both in terms of abnormal electrophoretic properties and heritable variations in catalytic activity [115-117]. 6PGD deficiency in terms of total enzyme activity was first described in 1964 by Brewer and Dern in which 6PGD with a 50% decrease in activity was observed in a Negro female [118]. To date, only a few 6PGD and 6PGD deficient individuals have been identified [118-121]. The association of 6PGD deficiency and hemolytic anemia is not always established. However, chronic hemolytic anemia in association with 6PGD deficiency was observed in two French patients and one Spanish family [122, 123]. Additionally, it was described that the low level of GSH was found in those patients. GSH instability and increased MDA were reported in the Spanish patient, leading to more susceptibility to oxidative stress of erythrocytes.

6-Phosphogluconate Dehydrogenase and Cancer

It was described that pentose phosphate shunt activity is elevated in proliferating cells. PPP is the major pathway providing ribose 5-phosphate, a precursor of nucleic acid biosynthesis and NADPH to the cell. Therefore, it was expected that this pathway might play a crucial role in tumor metabolism and rapidly dividing cells. 6PGD is the second dehydrogenase enzyme in this pathway and thus, it could serve as a marker of cell proliferation. In fact, both dehydrogenases (G6PD and 6PGD) were studied in many types of cancer cell lines and, interestingly, it was found that the activities of both enzymes increased when compared with those in normal cell lines [124]. Moreover, histochemical studies on tissues of human, rats and mice have demonstrated a significant increased activity of 6PGD in proliferating cells of both benign and malignant tumors [124].

Following the work described by Chayen in which the activity of 6PGD was elevated in tumor cells, a first attempt was made to measure the activity of 6PGD in vaginal fluid using a biochemical technique and the result was promising [125]. Later, 6PGD was used in a screening test for carcinoma of the cervix in parallel with a cytological method. Though increased activity of 6PGD was found in many patients, the screen failed to detect carcinoma in two of three cases. Owing to the fact that 6PGD activity was also elevated under benign conditions, measurement of 6PGD enzyme activity was not suitable as a cervical carcinoma screen [126]. Histochemical studies have been used to stain both G6PD and 6PGD enzymes in cervical smears and more intense signals for G6PD and 6PGD were identified in smears from the uterine cervix than from normal cells [127]. Later, several enzymes including 6PGD were tested in vaginal fluid as diagnostic markers of cervical cancer. 6PGD seemed to be a good candidate marker for cervical cancer because 6PGD activity was detected in 91% of patients in the particle-free fraction [128]. Additionally, biochemical measurements were developed to detect 6PGD activity in samples of vaginal fluid from patients compared with normal individuals [129]. Though an increase in enzyme activity was observed, it was inconclusive because of the contamination from erythrocyte enzymes. When a preparative step was performed to ensure that only the activities of G6PD and 6PGD enzymes from cervical epithelial cells was measured, it was evident from biochemical methods that 6PGD activity was increased in cervix epithelial cells in cervical intraepithelial neoplasia compared with normal individuals [130].

An increased 6PGD activity was discovered in colorectal carcinoma when compared wtih adjacent unaffected mucosa, reflecting a higher growth rate of tumor tissue [131]. This suggested that the enzyme might play a role in the level of gene expression in the colon tumor cell. Additionally, experimentally induced colon cancer metastases in rat also showed higher activity of 6PGD together with increased activity of SDH [132]. Induction of colon cancer metastases has led to an expansion of proliferation rate and capacity for purine catabolism in liver cells.

Though increased activity of 6PGD has been reported in many cancers as mentioned above, the mechanism by which 6PGD is activated and its role in human cancers remains largely unknown. The role of 6PGD in cancer has been investigated in NSCLC [133]. The expression level of 6PGD was found to be upregulated in lung cancer samples compared with normal lung tissues. Interestingly, inhibition of 6PGD significantly inhibited cell proliferation *in vitro* and in mice. However, 6PGD knockdown did not affect the level of NADPH and ribose 5-phosphate production, indicating that growth inhibition was not the result of a functional defect in the PPP. Rather, it was found that G6PD knockdown cells maintained the level of NADPH by enhancing glutaminolysis, an alternative pathway for NADPH production. Additionally, it was noticed that 6PGD knockdown cells were enlarged and showed the characteristics of senescence. Therefore, it was investigated whether inhibition of cell proliferation in 6PGD knockdown cells occurred through a senescence mechanism. Three markers of senescence, SA-β-GAL, p53 and p21, were tested and found to all be upregulated in tumor cells, indicating that inhibition of tumor growth occurred via a senescence-associated pathway. It was also revealed that inhibition of 6PGD resulted in a dramatic increase of metabolites in the oxidative branch of the PPP, 6-phospho-D-gluconate and D-gluconolactone-6-phosphate. This suggested that these two metabolites might be inhibitors of cell growth and 6PGD could be an attractive target for the treatment of NSCLC. A further study has shown that expression of 6PGD was associated with stage of lung carcinoma [134]. However, the correlation between 6PGD expression and lung cancer characteristics such as carcinoma type, tumor grade, and metastasis status remain unknown. Furthermore, inhibition of 6PGD significantly hindered phosphorylation at tyrosine 1234/1235 of c-Met, a hepatocyte growth factor receptor. Because tyrosine phosphorylation is crucial for the kinase activity of c-Met, inactivation of c-Met by knockdown of 6PGD led to the blocking of cell migration *in vitro*, which could result in growth inhibition of lung cancer cells.

A mechanism of 6PGD activity upregulation in cancer cells was illustrated in later study [135]. Proteomic-based studies have revealed that 6PGD was a target of lysine acetylation. Therefore, the effect of lysine acetylation on 6PGD activity was examined in different types of cancer cells such as lung, breast, head and neck and leukemia by treating these cells with inhibitors of the deacetylase enzyme. After treatment, the endogenous activity of 6PGD and levels of lysine acetylation were elevated, suggesting that acetylation on lysine residues of 6PGD activates the enzyme activity in human cancers. This was confirmed by mutagenesis study using acetyl-deficient mutants in which it was demonstrated that acetylation of both Lys76 and Lys294 is crucial for 6PGD activation. Structural analysis has shown that Lys76 is located adjacent to the cofactor binding site while Lys294 situated far from the catalytic site but might be involved in formation of the dimerization interface between two monomer molecules. Therefore, the role of acetylation of Lys76 and Lys294 was investigated and it was found that acetylation at Lys76 enhances binding of 6PGD to the $NADP^+$ cofactor while acetylation at Lys294 is important for dimer formation in 6PGD. Additionally, it was demonstrated in 6PGD-knockdown cancer cells that the enzyme is important for tumor growth in which acetylation of both Lys76 and Lys294 plays a crucial role and is required for proliferation and tumor cell growth. Also, acetyl CoA acetyltransferase 2 and dihydrolipoamide S-acetyltransferase (DLAT) were identified as upstream acetyltransferases and histone deacetylase 4 as a deacetylase of 6PGD. These results suggested that 6PGD might be a potential therapeutic target for cancer treatment.

It was shown that the results varied depending on the type of lung cancer cells studied. When H1975 lung cancer cells were used to study the functional role of 6PGD, it was found that knockdown of 6PGD did not alter the level of ribose 5-phosphate and NADPH in cells [133, 134]. However, when H1299 cells were used, suppression of 6PGD resulted in decreased oxidative PPP flux, ribose 5-phosphate production and NADPH [135, 136]. This discrepancy was suggested to be the result of different 6PGD functions and variations in the oncogenic background of different cancer cell types.

A study in H1299 cancer cells has revealed that 6PGD is important for the oxidative PPP and also contributes to proliferation and tumor growth of cancer cells [136]. This study also provided the information that 6PGD is important for regulating and coordinating the PPP in association with glycolysis in cancer cells. It was shown that 6PGD is involved in the activation of lipogenesis in a ribulose 5-phosphate-dependent mechanism in which ribulose 5-phosphate inhibits liver kinase B1-5' AMP-activated protein kinase (AMPK) signaling and subsequently activates acetyl-CoA carboxylase 1. Suppression of 6PGD resulted in reduced ribulose 5-phosphate production, increased AMPK activation and reduced cell proliferation.

Altogether, it was demonstrated that 6PGD plays a crucial role in proliferation and tumor growth in cancer cells. 6PGD links oxidative PPP to lipogenesis through ribulose 5-phosphate and acetylation of Lys76 and Lys294 of 6PGD is crucial for its activity. Additionally, 6PGD is necessary for cell migration via phosphorylation of c-Met. Therefore, 6PGD serves as an attractive target for cancer treatment, Figure 16.

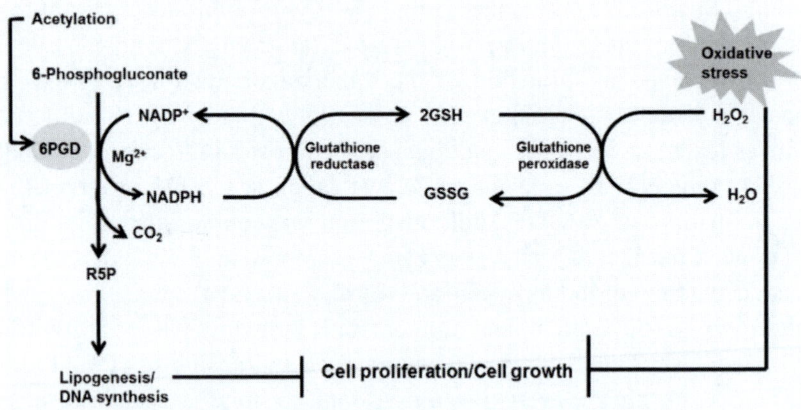

Figure 16. The role of 6PGD in cancer. The activity of 6PGD is regulated by acetylation at Lys76 and Lys294. Expression levels of 6PGD are upregulated in several cancers. The ribose-5-phosphate (R5P) produced in the PPP is necessary for the biosynthesis of lipid and DNA. Increased activity of 6PGD enhances the production of R5P, promoting cell growth and proliferation.

PYRUVATE DEHYDROGENASE COMPLEX

The pyruvate dehydrogenase complex (PDHc; EC 1.2.4.1) is a multienzyme complex that catalyzes the conversion of pyruvate into acetyl CoA, which then enters the TCA cycle (Figure 17). It comprises three enzymatic subunits: pyruvate dehydrogenase (E1), dihydrolipoamide acetyltransferase (E2), and dihydrolipoamide dehydrogenase (E3). The PDHc reaction is initiated by the oxidative decarboxylation of pyruvate by E1, which is considered to be rate-limiting step for the PDHc. During this process, an acetyl group derived from pyruvate decarboxylation is transferred to thiamine pyrophosphate. This acetyl group is then transferred to a lipoate moiety, which is covalently bound to E2. Next, E2 catalyzes transfer of the acetyl group from the lipoate moiety to CoA, resulting in the formation of acetyl CoA and dihydrolipoate. Dihydrolipoate is oxidized by E3 to regenerate lipoate that can participate in the next reaction cycle, while acetyl CoA then enters the TCA cycle for complete oxidative phosphorylation [137].

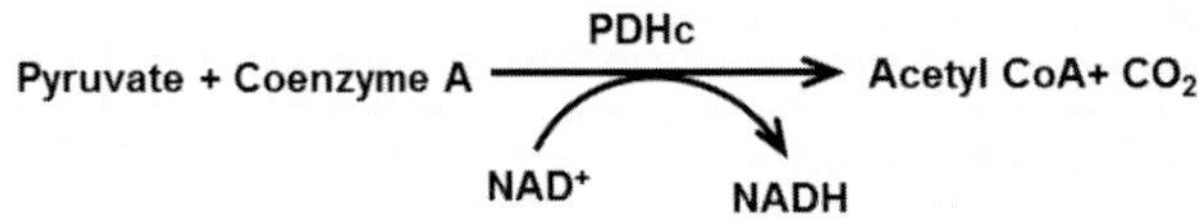

Figure 17. Reaction catalyzed by PDHc.

The function of PDHc is regulated by the phosphorylation status of E1 which is active when dephosphorylated. The phosphorylation of PDHc is performed by pyruvate dehydrogenase kinase (PDK), whose activity is activated by high levels of acetyl CoA, NADH, and ATP. In contrast, high levels of pyruvate and low energy state (high levels of NAD^+ and ADP) strongly inhibit PDK activity. Pyruvate dehydrogenase phosphatase (PDP) is involved in the dephosphorylation, and consequently reactivates the PDHc function. How PDP is regulated is unclear; however, Mg^{2+}, Ca^{2+}, and insulin have been shown to increase the activity of PDP [137, 138].

PDH-related diseases are most often caused by the loss of PDHc due to mutations in the E1 subunit. In addition to PDH subunit mutation, several case reports of patients with PDHc deficiency diseases demonstrated *PDP* gene mutation, without any mutation in PDHc. There is a wide spectrum of clinical presentations in PDHc deficiency depending on severity of the deficiency, mutation, and affected subunits. Patients generally present with lactic acidosis, which consequently contributes to various neurological and neuromuscular problems. In the absence of PDHc, pyruvate cannot be converted into acetyl CoA. The excess pyruvate is then converted to lactate and released into the bloodstream. Ultimately, changes in the acid-base balance of the blood and extracellular fluid alter normal cellular functions within the body.

Pyruvate Dehydrogenase and Cancer

A combination of proteomics and metabolomics was used to reveal the profile change of gastric cardia cancer (GCC), compared with corresponding non-cancerous gastric cardia tissue. It was found that the level of several enzymes and their related metabolic intermediates mainly involved in glucose metabolism were altered in GCC. Among these enzymes, lactate dehydrogenase A (LDHA) was upregulated and pyruvate dehydrogenase B (PDHB, also known as E1β subunit) was downregulated, and they were shown to be related to the elevated levels of lactic acid and pyruvic acid, respectively.

To assess the impact of PDHB in GCC development, further analysis performed *in vitro* by overexpressing PDHB. GCC cell lines Ooverexpressing PDHB were investigated for cell growth and metabolite profiling analysis. It was found that the stable overexpression of PDHB that was downregulated in GCC reduced lactic acid production, and impaired cell proliferation and colony formation on soft agar [139]. These data suggested that lactate formation, a favorable characteristic for cancer proliferation, is driven by the accumulation of pyruvic acid caused by the reduction of PDH. In contrast, the induction of PDH increased the conversion of pyruvic acid to acetyl CoA, resulting in a preference of the TCA cycle and reduced cancer survival. This study indicated that PDH might contribute to tumorigenesis and the Warburg effect.

The inhibitory effect of PDH on cancer progression was also revealed in other studies. PDH was immunohistochemically investigated in paraffin-embedded tissue from gastric cancer (GC) patients. PDH expression was much lower in cancerous tissues compared with normal mucosa specimens. Reduced PDH correlated with lymph node metastasis, lymphatic invasion, and histopathological grade of GC. In addition, PDH expression was positively correlated with the survival of GC, because patients with high levels of PDH had better overall survival [140]. Similarly, a study in surgically dissected ESCC supported these findings. The absence of PDHE1α protein expression in ESCC was found in 77% of cases, and was related to a poor histological grade of ESCC. Some cases of ESCC where follow-up information was obtained were enrolled in a survival analysis. Decreased PDHE1α protein expression in ESCC was significantly associated with a shorter overall survival [141].

Because the E1α subunit contains a catalytic site for pyruvic acid decarboxylation, the key reaction for the formation of acetyl CoA, studies of PDHc function often focus on PDHE1. However, some studies have reported associations of other PDH subunits with cancer development. An iTRAQ-based tandem mass spectrometry approach was employed to quantify the membrane proteome of GC cell lines. A difference in the expression of proteins between cancer and non-cancer samples was observed. Among these, DLAT (PDHE2), a novel cancer-associated protein, was upregulated in GC cell lines. The accumulation of pyruvic acid was observed along with decreased cancer cell proliferation following PDHE2 knockdown *in vitro*, indicating a role of PDHE2 in GC. Lack of the E2 subunit of PDHc resulted in blocking the conversion of pyruvic acid to acetyl CoA. However, in contrast to the Warburg effect, silencing PDHE2 decreased cancer growth [142]. The authors suggested that these GC cell lines may preferentially display a Reverse

Warburg effect. Not all cancers exhibit the Warburg effect, and peripheral cancer cells are likely to have access to a large O_2 supply, and therefore have the capacity to oxidize lactate into pyruvate and enter mitochondrial oxidative phosphorylation. This is termed the Reverse Warburg effect [143]. Taken together, these data suggest that PDHc adversely affects cancer development.

Normally, PDHc is thought to be an energy operator to support life. However, in cancer, PDHc is not beneficial to its progression and the negative functions of PDHc on tumorigenesis have been investigated. The transfection of HCC cell lines with a vector harboring *PDA1*, a gene encoding the E1α subunit, was performed. It was found that transduction of the *PDA1* vector induced significant late-stage apoptosis compared with the control plasmid. Interestingly, significant *PDA1*-induced cell death was not observed in the normal hepatocellular cell lines. To assess how the transduction of PDHE1α induced cancer cell apoptosis, regular apoptosis proteins involved in oxidative phosphorylation were examined. Western blotting analysis of *PDA1*-transfected cancer and normal cell extracts revealed the presence of Bcl-x/l. In contrast, Bcl-x/s was expressed only in the extracts of *PDA1*-transfected cancer cells. Bcl-x proteins are apoptotic regulators with two distinct isoforms. Bcl-x/l proteins act as apoptotic inhibitors while Bcl-x/s proteins act as pro-apoptotic proteins, involved in cytochrome c release. Cytochrome c release from mitochondria initiates caspase activation, a signaling pathway that participates in apoptotic cell death. Verification of these results by measuring cytochrome c demonstrated the translocation of cytochrome c into the cytoplasm of *PDA1*-transfected cancer cells, but not into normal cells or cells infected with a control vector [144]. In conclusion, PDH induces cancer cell death via caspase activation by initiating the release of mitochondrial cytochrome c. Consequently, this effect is only observed in cancer cells.

Other mechanisms were studied by McFate et al. Head and neck squamous cell carcinoma (HNSCC) cells lines showed decreased PDHc, and increased PDK-1 and phosphorylated E1α subunit. This suggested that a loss of PDHc occurs in cancer via the enhanced expression of PDK-1, resulting in the phosphorylation of PDHE1α into its inactive form. In addition, accumulated hypoxia-inducible factor-1α (HIF-1α), a transcriptional regulator of cellular responses to hypoxia, was also observed in HNSCC cells lines. To elucidate a possible role for PDHc inhibition in the phenotype of cancer cells, PDK-1 knockdown HNSCC cells lines were established by transfection with short hairpin (sh)RNA. The study revealed the reduction of phosphorylated PDHE1α and the induction of PDHc. Restored PDHc activity can reverse the Warburg metabolic phenotype as indicated by reduced lactate secretion into culture medium. Moreover, decreased HIF-1α was observed along with impaired cancer cell growth, inability to form colonies in soft agar, and reduced invasiveness [145]. Taken together, this study provided evidence that the absence of PDHc mediated by the overexpression of PDK caused the metabolic alteration, which in turn initiated HIF-1α stabilization. HIF-1α is malignancy-promoting transcription factor that upregulates the expression of hypoxic response genes, facilitating the aggression of cancer. Because the PDHc has inhibitory effects on tumorigenesis, PDHc is attracting increasing attention as a potential therapeutic target for cancer.

Pyruvate Dehydrogenase and Cardiovascular Disease

Although the majority of cardiac energy is derived from fatty acid oxidation, defects in PDHc, an enzyme in glucose metabolism, have been reported. Knockdown of the PDHE1α subunit in mice was used to investigate the role of PDHc in cardiac physiology. Homozygous E1α-knockdown mice grew normally during the preweaning period, but they did not survive for more than 7 days after weaning on a rodent chow diet. Although they survived on a high-fat diet, they still developed left ventricular hypertrophy and reduced left ventricular systolic function. This characteristic was also observed in heterozygous mice but with less severity. This mouse model demonstrated that a deficiency of PDHc caused defects in energy production from glucose metabolism, which in turn contributed to the cardiac pathologies [146].

Another related study was performed in transgenic mice overexpressing PDK-4. Heart tissue from PDK-4 transgenic mice showed increased PDK activity that corresponded with decreased PDHc activity. Reduced glucose oxidation was observed in heart extracts from PDK-4 transgenic mice along with enhanced fatty acid oxidation. These results indicated the presence of a compensatory mechanism for energy production by altering substrate utilization. Because the energy supply was maintained, no abnormalities or hypertrophy, or changes in left ventricular function were observed. However, introduction of the PDK-4 transgene into mice expressing a constitutively active form of the phosphatase calcineurin, which causes cardiac hypertrophy, induced cardiomyocyte fibrosis and a significant increase in mortality. In conclusion, loss of PDHc caused by the overexpression of PDK is sufficient to cause metabolic inflexibility, which in turn may exacerbate cardiomyopathy mediated by the calcineurin signaling pathway [147].

Because the heart has metabolic flexibility by adjusting energy production based on substance availability, the effect of PDHc on cardiomyopathies is not easily demonstrated in normal heart. Cardiac efficiency and the recovery of contractile function in post-ischemic hearts can be improved by PDHc. In a rabbit model study, ischemic hearts were modelled by injecting heparin. Then, hearts were perfused with pyruvate with or without dichloroacetate (DCA), an inhibitor of PDK, to maintain PDHc activity. It was found that post-ischemic hearts with augmented pyruvate oxidation (DCA-treated) sustained improved mechanical function throughout 40 minutes of reperfusion [148]. The benefit of PDHc activation by DCA for functional recovery in post-ischemia heart was also demonstrated in other studies [149, 150]. These evidences suggest that the sustained activation of PDHc results in the overproduction of energy, which is a beneficial mechanism for maintaining the regular function of the cardiac system.

Pyruvate Dehydrogenase and Neurodegenerative Disease

The brain is a crucial organ that requires a large energy supply. Glucose metabolism is the main pathway to generate energy for the human brain. Approximately 30% of the daily glucose consumption is used to provide energy for the brain. Therefore, altered glucose metabolism is a key factor in neurological pathogenesis. Many studies have reported that neurodegenerative diseases are caused by PDHc deficiency, which mostly present as a severe inborn error of metabolism.

A case of PDHc deficiency with antenatal brain atrophy was reported by Wada et al. A routine ultrasonographic study performed at the 28th week of gestation revealed enlargement of the lateral ventricles of the fetus, a characteristic indicating abnormal brain formation. At the 40th week of gestation, laboratory findings of this female infant indicated metabolic acidosis and elevated concentrations of lactic acid and pyruvic acid. A presumptive diagnosis of PDHc deficiency was made. A seven base-pair deletion (positions 927–940) in exon 10 of the PDHE1α subunit gene was observed, which created a termination codon downstream of the deleted segment. Evaluation at 6 months of age revealed an anomaly of facial physics and no psychomotor development [151].

Another patient with a different mutation was reported by Tamura et al. Fetal ultrasound and magnetic resonance imaging revealed that abnormal brain development was present at the 25th week of gestation. A presumptive diagnosis of PDHc deficiency was determined based on the heterozygous mutation c.1140_1150dup, p.I384TfsX43 in the E1α subunit gene. The female neonate had no external or visceral malformations except a high arched palate and brain malformation, and presented with psychomotor delay [152].

Antenatal neurometabolic disorder caused by PDHc deficiency has been reported. In some cases, severe symptoms have been reported in adults. A 20-year-old man with PDHc deficiency presented with worsening blurry vision in the right eye. He was diagnosed with PDHc deficiency at the age of 18 months, indicated by lactic acidosis and reduced PDHc activity. Between the age of diagnosis and time of the report, he did not show serious signs of disease, only muscle weakness, spasms, and mild developmental delay. He had diplopia until the age of 18 years. At 20 years of age, he presented with slight exotropia and poor color vision in the right eye. Magnetic resonance imaging at this time demonstrated abnormal signals in both optic nerves and in the optic chiasm [153].

Becasue various patterns of PDHc-deficient neurologic diseases have been reported, Barnerias et al. examined the phenotype and genotype of PDHc deficiency in 22 participants with enzymological and genetically confirmed PDHc deficiency by investigating clinical and imaging features over a 15-year period. They classified PDHc deficiency into four neurologic presentations: (1) early encephalopathy with brain malformations involving the corpus callosum and/or cortex; (2) acute brainstem dysfunction at infancy with or without the typical basal ganglia abnormalities of Leigh syndrome; (3) congenital motor disorders suggesting cerebral palsy but combined with neuropathy or paroxysmal dystonia; and (4) relapsing ataxia, a chronic axonal neuropathy with recurrent motor deficiency. Genotype-phonotype correlations showed that most mutations occurred in the PDHE1α subunit gene, of which mutation in exons 6, 7, and 8 are associated with Leigh syndrome and relapsing ataxia. Mutations in other PDHx genes were correlated with non-progressive encephalopathy and long-term survival [154]. Based on this study, the correlation of neurological and survival outcomes in PDHc deficiency with genotype have been established. Nevertheless, a later study performed by DeBrosse et al. reported a contrary result. The clinical outcomes and neurological features of 59 participants with confirmed PDHc deficiency and defined mutations in one of the PDHc genes were analyzed. The outcomes of this population with genetically confirmed PDC deficiency were heterogeneous and not distinctive [155].

Because there are many factors involved in brain glucose metabolism, variations in the clinical pattern of neurologic morbidity in PDHc deficiency might occur, resulting in the conflict of these two studies. To investigate the underlying molecular mechanism, unpredictable factors should be considered. A study of PDHc function in neurodegenerative disease was performed in a rat model. Rats with traumatic brain injury (TBI) were induced and then treated with pyruvate, a substrate for the PDHc enzyme. At 72 h post injury, blood and brain tissue from TBI induced rats with or without pyruvate treatment were analyzed. Reduced PDHc enzyme was observed in TBI along with an increased lactate to pyruvate ratio, decreased base excess, and increased malondialdehyde, a marker of oxidative stress. Lower PDHc in the blood was related to increased gliosis and the loss of PDHE1α subunit in brain tissue. These results indicated that TBI causes systemic metabolic alterations in the brain by introducing metabolic acidosis. Moreover, alterations in PDHc enzyme levels may influence the outcomes of TBI. Interestingly, the administration of pyruvate was significantly effective in reversing these harmful metabolic pathways, leading to the reduction of gliosis and decreased loss of PDHc [156]. Although this study did not directly show a role for PDH in brain function, it provided supporting evidence that metabolic flux into glucose oxidation, usually driven by PDHc, is protective against brain injury.

Further evidence was provided using a mouse model in which the gene encoding the α subunit of PDHc was interrupted. Reduced PDHc activity verified the success of silencing the gene. The oxidation of glucose to CO_2 and incorporation of glucose-carbon into fatty acids were reduced in the brain of 15 day-old PDC-deficient mice. Histological analyses showed alterations in several structures in the white and gray matter in 35-day-old PDC-deficient mice. Furthermore, cell proliferation, migration, and differentiation into neurons by newly generated cells were reduced. Locomotor analysis revealed normal activity in a novel environment but a decreased startle response to loud noises. They also exhibited abnormal pre-pulse inhibition of the startle reflex. Taken together, this study provided excellent evidence that a reduction of glucose metabolism and fatty acid synthesis caused by a loss of PDHc resulted in the impairment of neuron proliferation, differentiation, as well as brain development [157].

Pyruvate Dehydrogenase and Diabetes

Because the mitochondrial PDHc controls the rate of glucose oxidation, impaired PDHc activity is associated with insulin resistance or DM. A murine model of β-cell PDHc deficiency (β-PDHKO) revealed decreased pancreatic insulin. The plasma insulin level was also decreased along with increased blood glucose levels. The induction of islets isolated from β-PDHKO mice with glucose did not activate insulin secretion. In addition, impaired glucose tolerance and *in vivo* insulin secretion during hyperglycemic clamp were evident in β-PDHKO mice. However, no change in the number or size of islets was observed in the pancreas of these mice, indicating β-cell damage had not occurred. Consequently, this β-PDHKO mouse model demonstrated that a loss of PDHc enzyme altered glucose-stimulated insulin secretion by β-cells, which in turn may contribute to hyperglycemia [158].

The glucose-fatty acid cycle was proposed by Randle et al. in 1963. It describes the competition of glucose and fatty acid for energy production substrates. Because fatty acid is the preferred energy source for respiration, high-fat intake inhibits glucose utilization by controlling PDHc activity. This has been considered a causative factor for insulin resistance and type 2 DM. Growth hormones (GH) were reported to cause pronounced skeletal muscle insulin resistance. Because GH stimulates lipolysis, it was hypothesized that the suppression of PDHc activity might underlie GH-induced insulin resistance. It was found that fasting and GH promoted lipolysis, which was associated with a 50% reduction in insulin sensitivity compared with healthy controls. Reduced PDHc activity was observed along with increased phosphorylated PDHE1α and PDK-4 mRNA. Therefore, this study suggested that the inactivation of PDHc is an underlying mechanism of GH-induced insulin resistance [159].

Furthermore, another study reported the pivotal role of PDHc in DM using PDK-4 knockdown mice fed a high-fat diet. Insulin resistance developed in both wild-type and PDK-4 knockdown mice. Nevertheless, fasting blood glucose levels were lower, glucose tolerance was slightly improved, and insulin sensitivity was slightly greater in the PDK-4 knockdown mice compared with wild-type. These results demonstrated that maintaining PDHc function by removing its inhibitor can reduce the risk and severity of glucose tolerance in diet-induced obesity [160].

Diabetic ketoacidosis (DKA) is a life-threatening complication of DM. This condition occurs when the body cannot use glucose as an energy source, because there is insufficient insulin. Therefore, fatty acids are broken down and harmful acidic ketone bodies are produced. Andersen et al. reported that patients presenting with DKA have lower PDHc activity and quantities compared with healthy controls and hyperglycemic patients without ketoacidosis. PDHc activity was also significantly correlated with serum bicarbonate and pH and inversely correlated with the anion gap [161].

However, further studies to determine whether PDHc depression leads to DKA development or whether DKA decreases the expression of PDHc is required.

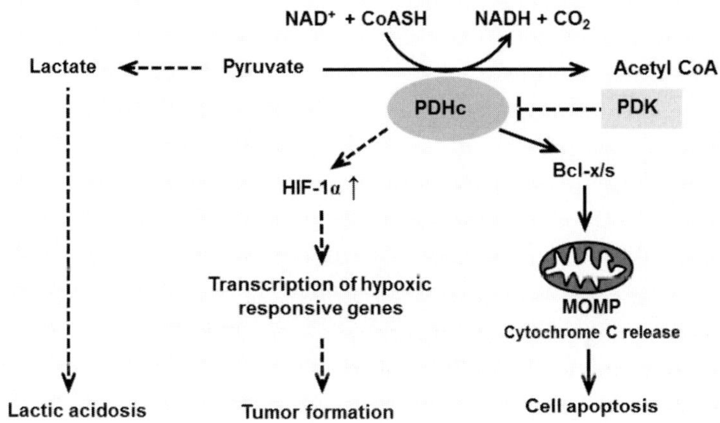

Figure 18. The role of PDHc in disease development. Normally, PDHc regulates programmed cell death by controlling the levels of apoptotic signaling molecules. Loss of PDHc causes the defect in HIF-1α degradation. HIF-1α signaling activates the expressions of several hypoxic responsive genes, enhancing tumorigenesis. PDHc deficiency results in pyruvate shunt to lactate which may lead to lactic acidosis. Pyruvate dehydrogenase kinase (PDK) inactivates PDHc activity.

Collectively, PDHc plays an important role in several diseases such as cancer, cardiovascular and neurodegenerative diseases. PDHc is downrugulated in many cancer cells, promoting the Warburg effect. PDHc induces cell death via caspase activation. Mutations and deficiency of PDHc cause a reduction in its activity, altering glucose metabolism. This leads to an aberration of neuron proliferation, differentiation and brain development. The role of PDHc in disease development is demonstrated in Figure 18.

LACTATE DEHYDROGENASE

Lactate dehydrogenase (LDH; EC 1.1.1.27) is an enzyme catalyzing the reversible transformation of pyruvate to lactate under anaerobic conditions with concomitant conversion of NADH to NAD^+ (Figure 19). Active LDH is a homo- or hetero-tetrameric enzyme consisting of two subunits (LDHA and LDHB), each with a molecular weight of 35 kDa. These two separate subunits are encoded by separate genes which are highly related, *LDHA* and *LDHB*. In fact, there are four LDH genes, namely *LDHA, LDHB, LDHC* and *LDHD* in which *LDHA, LDHB,* and *LDHC* are the L isomers and *LDHD* is the D isomer. The L isomers use and produce L-lactate, commonly found in vertebrates, whereas the D isomer uses and produces D-lactate. *LDHA* and *LDHB* are believed to have arisen from duplication of a single LDHA-like LDH gene [162]. *LDHC* is a testes-specific gene which is thought to have evolved from duplication of the *LDHA* gene after A-B duplication in mammals.

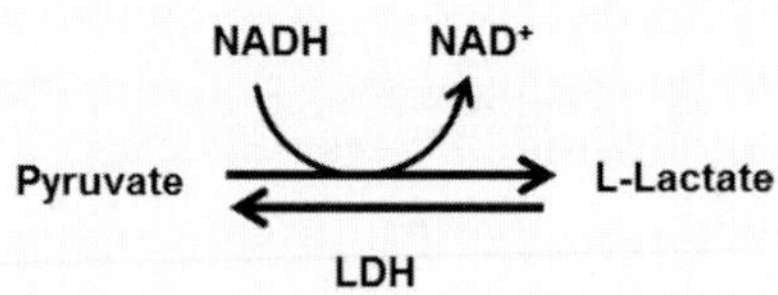

Figure 19. Reaction catalyzed by LDH.

LDHA is also known as subunit M as it is highly expressed in skeletal muscle and LDHB is known as subunit H because it is predominantly found in the heart. While LDHA and LDHB can form both homo- or hetero-tetramers, other LDHs can produce only homo-tetrameric enzyme. There are five LDH isoforms which can be formed from the M and H subunits, LDH1 (4H), LDH2 (3H, 1M), LDH3 (2H, 2M), LDH4 (1H, 3M), and LDH5 (5M). LDHC is also referred to as LDH6 or LDHX. LDHA preferentially catalyzes the transformation of pyruvate to lactate while LDHB catalyzes the conversion of lactate to pyruvate. Therefore, the composition of LDHA and LDHB in each LDH isoform determines the ability of the enzyme to catalyze the reaction. Each LDH isoform shows different electrophoretic mobility on a polyacrylamide gel in which LDH5 is the slowest migrating isoform.

Lactate Dehydrogenase and Cancer

The hallmarks of cancer include the metabolic alterations which allow cancer cells to rapidly proliferate in a hostile environment. One that has been extensively investigated is aerobic glycolysis, also known as Warburg effect. Aerobic glycolysis and lactate formation were found to promote cell proliferation, cell growth and invasion/migration in cancer cells. Additionally, the importance of aerobic glycolysis in cancer initiation, development and progression has been recognized. Mechanisms underlying these phenotypes and their functional significance in cancers have been extensively investigated. Evidence has indicated that LDH plays a critical role in tumorigenesis. High expression of LDH was identified in many cancer types and found to be mediated by a number of oncogenes.

In pancreatic cancer, expression of LDHA at the mRNA and protein levels was increased in clinical pancreatic cancer samples [163]. Overexpression of LDHA enhanced cell growth, inducing S phase transition and increasing colony formation. However, knockdown of LDHA resulted in cell growth inhibition. The apoptosis pathway was activated as shown by an increase in expression of poly ADP ribose polymerase, an indicator of cell apoptosis, and decreased expression of the anti-apoptotic genes, X-linked inhibitor of apoptosis protein, Bcl-2, and Bcl-x/l. Additionally, suppression of LDHA expression inhibited the tumorigenicity of pancreatic cancer cells in a xenograft mouse model. The results suggested an oncogenic role of LDHA in cell growth and progression of pancreatic cancer. Further evidence was provided to support the oncogenic role of LDHA. Acetylation of LDHA was observed at Lys5, causing reduction in the catalytic activity and protein expression levels. When LDHA was replaced with an acetylation mimic mutant, cellular proliferation and migration were decreased, suggesting the role of LDHA Lys5 acetylation in the regulation of cell growth. Early and late stage pancreatic cancers exhibited decreased Lys5 acetylation in association with increased expression of LDHA, suggesting the contribution of Lys5 acetylation to pancreatic cancer initiation [164].

Furthermore, expression of LDHA was found to be regulated by many transcription factors, which likely contributes to proliferation, invasion, progression and development of pancreatic cancer. Expression of LDHA was negatively correlated with Kruppel-like factor 4 (KLF4), a transcription factor and putative tumor suppressor, in pancreatic cancer [165]. Expression of LDHA was increased in KLF4 null MEFs cells and, among glycolysis-related genes, LDHA was the most affected of those under the regulation of KLF4. Overexpression of LDHA was correlated with disease stage and tumor differentiation. Upregulation of LDHA was found in late stage pancreatic cancer and was correlated with decreased cell differentiation, suggesting a key role of LDHA in tumor development and differentiation. KLF4 was found to bind the promoter region of LDHA and inhibited its transcription, suggesting that the KLF4/LDHA pathway plays a critical role in aerobic glycolysis and development and progression of pancreatic cancer. LDHA was also found to correlate with the transcription factor forkhead box protein M1 (FOXM1) [166]. FOXM1 expression promotes the Warburg effect in pancreatic tumors, increasing aerobic glycolysis. Increased expression of LDHA was accompanied by overexpression of FOXM1 in pancreatic tumors and cancer cell lines. In fact, it was found that expression of FOXM1 upregulated the expression of LDHA at both the mRNA and protein levels by binding to the LDHA promoter region, increasing LDHA activity, lactate production and utilization of glucose. Enhanced FOXM1/LDHA signaling was also found to increase metastasis and cell growth *in vivo*, contributing to the development and progression of pancreatic cancer.

Increased expression of LDHA in pancreatic cancer was also found to be associated with HIF-1α, a transcription factor response to hypoxia. Under hypoxia, expression of HIF-1α and production of lactate were elevated [167], leading to an increase of glycolysis in pancreatic cancer cells. This also enhanced *in vitro* cell proliferation and decreased apoptosis. Furthermore, association between c-Myc and LDHA was found to affect the progression of pancreatic cancer [168]. Overexpression of c-Myc, a regulator gene encoding for transcription factor, was concomitant with increased expression of LDHA in pancreatic cancer samples and cell lines. Regulation of LDHA by c-Myc promotes the Warburg effect, increasing aerobic glycolysis. Expression of both c-Myc and LDHA was correlated with tumor size and tumor, nodes, metastasis (TNM) stage. Additionally, overexpression of c-Myc and LDHA was associated with poor prognosis in pancreatic cancer patients.

In addition to LDHA, LDHB was found to play a role in pancreatic cancer development and progression. It was reported that suppression of LDHB promotes pancreatic cancer progression via induction of the glycolytic phenotype [169]. In contrast to LDHA, LDHB expression was decreased and even lost in pancreatic cancer cells. No association was observed between expression of LDHB and sex, age, regional lymph node metastatis or differentiation of pancreatic cancer cells. Hypermethylation at the promoter region of LDHB was found to contribute to decreased expression of LDHB in pancreatic cancer cells, increasing LDH activity which led to enhancement of lactate production and decreased ATP formation. High expression of LDHB was found to inhibit cell proliferation whereas knockdown of LDHB resulted in increased cell proliferation under hypoxic conditions. Furthermore, expression of LDHB decreased the flattening, spreading, migration and invasion of pancreatic cancer cells under hypoxia.

In breast cancer, expression of the LDH isoforms LDHA and LDHB was investigated [170]. LDHA and LDHB were expressed in all breast cell lines used in this study, two breast cancer cell lines, MCF-7 and MDA-MB231, and a control breast cell line, HMEC 184. LDHA was predominantly expressed in MDA-MB231 and HMEC 184 cells. On the other hand, LDHB was highly expressed in MCF-7 cells. The serum level of LDH was also found to be increased, indicating high clinical TNM staging [171]. An elevated serum level of LDH indicated poor prognosis, surgical outcome and metastasis in breast cancer patients. The serum level of LDH was also proposed as a potential prognostic marker for overall survival in non-metastatic breast cancer [172]. LDH levels were significantly correlated with survival in a retrospective study of bone metastases from breast cancer patients [173].

To specifically investigate the role of the LDHA isoform in breast cancer, LDHA expression was inhibited using shRNA in the breast cancer cell lines MCF-7 and MDA-MB231 [174]. Expression of LDHA was found to be strongly associated with tumor size. Both *in vivo* and *in vitro* results revealed that silencing LDHA inhibited cell proliferation, increased intracellular ROS and induced mitochondrial pathway apoptosis. Moreover, knockdown of LDHA decreased tumorigenesis in breast cancer xenografts. Additional data also confirmed the importance of LDHA in metastatic breast cancer [175]. This was consistent with a previous report where the inhibition of LDHA, also by shRNA, caused reductions in cell proliferation, cell invasion and migration, glucose consumption, glycolysis and lactate production, and increases in the consumption of oxygen, production of ROS and cellular ATP levels. The data indicated the LDHA isoform was the tumorigenic agent in breast cancer. A further study indicated the functional role of LDHA in breast cancer [176]. In agreement with previous data, LDHA was found to be upregulated in breast cancer cell lines and clinical samples. Increased expression of LDHA was correlated with cell proliferation, metastasis, poor overall survival and disease free survival in breast cancer patients. Additionally, *in vivo* and *in vitro* experiments indicated that LDHA increased glycolysis and cell proliferation. Importantly, it was discovered that LDHA was a direct target of the miR-34a family, a group of non-coding RNA molecules. miR-34a was identified as a tumor suppressor. In breast cancer, miR-34a exerts its function by targeting the 3'UTR of LDHA. Repression of LDHA by miR-34a resulted in decreased glycolysis and cell proliferation in breast cancer. High expression of LDHA might be accompanied by the loss of LDHB. Silencing of LDHB in breast cancer has been reported [177]. It has been revealed that loss of LDHB occurred at an early (DCIS) stage and this loss was caused by methylation in the promoter region of LDHB. Suppression of LDHB resulted in a marked increase in lactate production in hypoxia, promoting cell growth and survival.

However, this study indicated that LDHA was highly expressed in MCF-7 in contrast to the LDHA expression level observed by Hussien and Brooks [170].

When a pharmacological inhibitor of NF-κB, panepoxydone (PP), was used in breast cancer, the expression of both LDHA and LDHB proteins could be regulated [178]. PP inhibited expression of LDHA and increased expression of LDHB in the breast cancer cell lines MCF-7, MDA-MB231, MDA-MB468 and MDA-MB453. PP also inhibited oxygen consumption, lactate production, ATP synthesis and mitochondria membrane potential. Increased expression of LDHB by PP promoted apoptosis and mitochondrial respiratory damage while diminishing cell migration. Interestingly, analysis of functional genomic data (GEO) revealed that 100% of non-triple negative breast cancers and 60% of triple negative breast cancers showed less expression of LDHB than LDHA.

LDHB was also found to play an important role in breast cancer. An integrative analysis of triple negative breast cancers indicated the importance of LDHB for cell proliferation and tumor growth [179]. LDH expression was elevated and found to act as a regulator in basal-like/triple negative breast cancers. Knockdown of LDHB by siRNA caused a reduction in proliferation of triple negative breast cancer cell lines. Additionally, xenograft tumor growth was repressed when LDHB was inhibited. Microarray analysis also showed upregulation of glycolytic genes and downregulation of oxidative phosphorylation genes in triple negative breast cancers and cell lines, indicating the dependence of breast cancer on glycolytic metabolism. Furthermore, poor clinical outcomes were associated with expression of LDHB. In support of this notion, another study reported the increased expression of LDHB in breast cancer cell lines with glycolytic, basal-like phenotypes [180]. High expression of LDHB predicted the response of neoadjuvant chemotherapy in breast cancer patients. As the expression of LDHA and LDHB seemed to vary among breast cancer cell lines, it was proposed that rather than focus on the expression of only one protein, the ratio of LDHB/LDHA or the glycolytic index would be more useful to detect oxidative phosphorylation in breast cancer cells [178].

The association between LDHB and cancer cell invasiveness was also elucidated. Increased glycolytic lactate production with defective mitochondrial respiration was observed in hepatoma cells and HCC tissues [181]. The invasive activity of hepatoma cells was found to be mediated by claudin-1 (Cln-1), a tight junction protein, via mitochondrial respiratory defects. Increased lactate production was caused by down-regulation of LDHB rather than LDHA. Suppression of LDHB expression activated Cln-1 at the transcription level as well as impaired mitochondrial respiratory activity, enhancing the invasiveness of Chang cells. The results indicated the regulatory role of LDHB in stimulating mitochondrial respiratory defects and its contribution to cancer cell invasiveness by inducing Cln-1.

In thyroid cancer, expression of LDH in human thyroid tissues and thyroid cell lines was examined [182]. It was observed that the LDHA/LDHB expression ratio was lower in follicular and oncocytic tumors than normal thyroid tissues. LDHB expression was associated with estrogen-related receptor α (ERRα) whereby ERRα interacts with the LDHB promoter and exerts a negative transcriptional activity, decreasing LDHB expression at the mRNA and protein levels. Therefore, it was proposed that thyroid cancer cells reprogram the metabolic pathways via increased expression of ERRα, decreasing LDHB activity. Reduced expression of LDHB will change the subunit composition of LDH, transforming the LDH isoform into LDH5 which contains only the LDHA gene. Importantly, LDHA is known to associate with metastatic progression [183]. Therefore, ERRα could increase aggressiveness by regulating the LDHA/LDHB ratio in thyroid tumors. Accordingly, another study also addressed increased expression and phosphorylation at Tyr10 of LDHA in follicular and papillary thyroid cancers [184]. It was also revealed that phosphorylation of LDHA occurred in a fibroblast growth factor receptor 1 (FGFR1)-specific manner. Inhibition of FGFR1 by dovitinib and SU-5402 significantly decreased phosphorylation of LDHA.

In GC patients with different grades and stages of the disease, expression of LDH5 (LDHA) was examined [185]. High expression of LDH5 together with increased expression of HIF-1α was observed in approximately 62% of the patients. A strong association was observed for expression of LDH5, HIF-1α and two angiogenesis factors (VEGF and cyclooxygenase-2 (COX-2)). Increased expression of LDH5 was found to be predominant in patients with advanced tumors and vessel invasion. Under hypoxic conditions, expression of HIF-1α induced glucose metabolism which is advantageous for the proliferation as well as aggressiveness of cancer cells. Furthermore, overexpression of LDH5 was proposed as a marker for poor prognosis in GC. Serum LDH levels in GC stric cancer were also determined [186]. A strong association was observed between serum LDH levels and pathological T stage, metastasis, TNM stage and recurrence of GC. In gastric patients, increased serum LDH levels likely indicated shorter overall and disease-free survival rates, suggesting that LDH could be a poor prognostic marker in GC.

Taken together, LDH is overexpressed in many cancers, indicating that it plays a critical role in cancer. LDH produces lactate which is essential for cell growth, proliferation and survival, enhancing the Warburg effect which is beneficial for cancer cells. The role of LDH in cancer development is illustrated in Figure 20. Overexpression of LDH is regulated by transcription factors such as FOXM1, HIF-1α and c-Myc.

A further study revealed the mechanism by which LDH is inhibited by oxamate [187]. Increased expression of LDH in GC was also observed in the same study. High expression of LDH was found to be strongly associated with the pathological TNM stage, carcinoembryonic antigen and poor prognosis, and was consistent with a previous report. It was demonstrated that inhibition of LDH by oxamate resulted in reduction of cell viability as well as induced protective autophagy in GC. Additionally, inhibition of LDH also reduced cell proliferation and increased ROS production. Inhibition of LDH by oxamate was found to be mediated by the protein kinase B (Akt)–mTOR signaling pathway, a negative regulator of autophagy. It was previously reported that LDHB was upregulated in an mTOR-dependent manner in activated Akt1-expressing mouse embryonic fibroblast [188]. This indicated that LDHB is a downstream target of mTOR, a protein that is important in tumorigenesis.

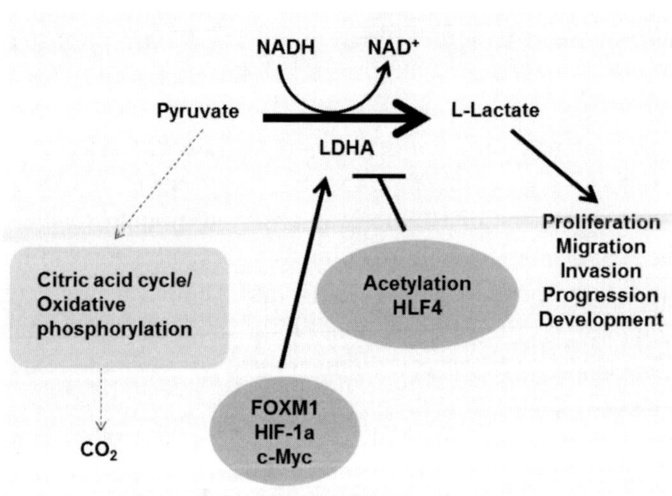

Figure 20. The role of LDH in cancer. Lactate, the product of LDH reaction, is required for cell proliferation, migration and invasion. The activity of LDH is regulated by several factors. Acetylation at Lys5 and transcription factor KLF 4 inhibit the expression of LDH. Overexpression of LDH is observed in many types of cancer. Transcription factors such as FOXM1, HIF-1α and c-Myc increase the expression of LDH in cancer cells, promoting cell growth, proliferation and invasion which contribute to the development and progression of tumor cells.

LDH is abundantly expressed in most of the tissues in the body, including the kidney, heart, skeletal muscle, pancreas, spleen, liver and lung. Since normal cells do not secret LDH, increased LDH levels can be used as a marker of tissue damage. Elevated LDH levels were observed as a tissue injury marker in the heart, blood vessels, liver, brain, kidney and skeletal muscle. It was believed that malignant tumor tissue or damaged tissue caused by cancer resulted in the release of LDH into circulation, elevating enzyme levels. Generally, damaged cells are removed by phagocytosis by phagocytes that are activated by substances released from damaged cells. It was supposed that high levels of LDH leaked from damaged tissues might contribute to phagocytosis activation. LDH was found to increase the production of interleukin-6 (IL-6) and TNF-α in a dose-dependent fashion in the macrophage like cell line J774.1 via increasing C-Jun NH_2-terminal kinase (JNK) phosphorylation and NF-κB translocation to the nucleus [189]. Additionally, it was also observed that transcription levels of both IL-6 and TNF-α were elevated. These results indicated the ability of LDH to activate macrophages.

In addition to cancer, LDH also plays a role in osteoclast differentiation. It was concluded that glycolytic metabolism was elevated during osteoclast differentiation when glucose consumption and lactate production were increased [190]. Oxygen consumption was also found to increase during differentiation. Suppression of LDHA or LDHB resulted in decreased glycolytic and mitochondria metabolisms, causing decreased mature osteoclast formation. It was observed that this occurred through reduced osteoclast precursor fusion and downregulation of osteoclastogenic critical transcription factor, nuclear factor of activated T-cell 1 (NFATc1), as well as its target genes, suggesting that receptor activation during NF-κB ligand (RANKL)-induced LDH activation promotes glycolytic and mitochondria metabolisms. Increased expression and activity of LDH during osteoclast differentiation promotes glycolytic and mitochondria metabolism, enhancing the transformation from osteoclast precursors to mature osteoclasts. Therefore, LDH could be a potential therapeutic target for treatment of osteoporotic disease.

Lactate Dehydrogenase as a Prognosis Biomarker

High LDH levels were observed in several cancers. Therefore, LDH was proposed to be a potential prognosis marker for cancers such as lymphomas, pancreatic cancer, RCC, GC, bladder cancer and melanoma [172, 191-195]. It was believed that increased expression of LDHA is associated with poor prognosis and resistance to chemotherapy and radiotherapy in cancers. Though this notion was proven for several cancers, it was not applicable to certain cancers. Increased LDH was reported to serve as a poor prognosis marker of primary gastric diffuse large B cell lymphoma in a retrospective study of 46 cases of the Hong Kong-Chinese population [191]. This observation was further confirmed from a larger sample size in which high serum LDH levels showed significant prognostic impact for diffuse large B cell lymphoma in Japanese patients [196].

Furthermore, LDHA was also found to be markedly expressed in gastric carcinoma. A positive correlation was observed between LDHA expression and age, histological type and lymph node metastases. High expression of LDHA was associated with poor overall survival [194, 197]. Increase in LDH5 (LDHA) expression was correlated with disease progression in melanoma, indicating its prognostic significance [198]. LDH5 expression was closely related with increased tumor thickness and derma tumor mitotic rate. High expression of LDH5 i.e., above the median immunoreactive score was found to be negatively correlated with disease free survival and overall survival. Multivariate analysis indicated that LDH5 was not an independent prognostic marker for melanoma. However, LDH5 expression could be used as a marker for the degree of malignancy. LDH5 expression was found to reduce the expression of Bcl-2, an anti-apoptotic protein. Nonetheless, the expression of LDH5 was associated with the expression of two other anti-apoptotic proteins, Bcl-x/l and Mcl-1, and the endoplasmic reticulum stress protein glucose-regulated protein 78.

Metastatic RCC patients exhibited high levels of LDH. Patients treated with VEGF-targeted therapy showed low LDH levels with an improved possibility of treatment outcome when compared with patients with high levels of LDH [193]. The overall survival of patients was significantly associated with levels of LDH, indicating the possibility of LHD as a potential prognostic marker. It was also suggested that levels of LDH could be a predictive marker for metastatic RCC treated with VEGF signaling inhibitors. Additionally, treatment with anti-VEGF antibody, bevacizumab, in combination with chemotherapy decreased the progressive rate and improved the progressive free survival in colorectal cancer patients with high LDH levels [199].

In pancreatic cancer, a differential expression pattern of LDHA was observed [200]. Expression of LDHA was relatively low at a very early stage and increased along the carcinogenetic pathway from pancreatitis through cyst/PanIN to cancer and the highest expression was found in the most aggressive tumors. Overexpression of LDHA was significantly associated with tumor cell proliferation, migration, invasion and angiogenesis. Furthermore, increased expression of LDHA was also found to contribute to the aggressiveness of cancer. Multivariable analysis showed that expression of LDHA was a poor prognostic marker for survival in pancreatic cancer patients In line with a previous study, it was also shown that LDH was a valuable prognostic and predictive marker for advanced pancreatic cancer patients treated with sorafenib [201]. Though it was found that acetylation of Lys5 was decreased in pancreatic cancer no correlation was found between reduced Lys5 acetylation and liver cancer initiation, limiting the use of LDHA as a biomarker for certain cancers such as pancreatic cancer but not for other cancers such as liver cancer [164]. Lys5 acetylation of LDHA can be readily detected by a specific antibody; therefore, it might also serve as a diagnosis marker for some cancers. LDHA could be a valuable therapeutic target because of its important role in the maintenance of glycolysis in cancer cells. Direct inhibition of LDHA, LDHS acetyltransferase or LDHS deacetylase may provide a therapeutic strategy for cancers.

In contrast, serum LDH was not correlated with treatment (surgery, radiotherapy, and chemotherapy or system treatment) response in stage IV melanoma, indicating that LDH had no prognostic utility in stage IV melanoma [192]. However, serum LDH levels were significantly associated with the stage of disease. The levels of LDH were significantly higher in stage IV melanoma when compared with the control group of stage I/II. However, the serum levels of LDH were not useful to differentiate between healthy controls and melanoma patients in the early stage of disease.

Lactate Dehydrogenase as a Diagnostic Marker

Analogous to other cancers, aerobic glycolysis is essential for cell proliferation and migration in oral squamous cell carcinoma (OSCC). It was observed that expression of LDH5 or the salivary level of LDH is increased in OSCC [202, 203]. As saliva LDH was released from oral epithelial cells, alteration in the levels of LDH5 in saliva should indicate the metabolic or pathological changes in these cells, reflecting the biological behavior or tissue damage of OSCC [203]. Increased tissue expression of LDH5 was found to be correlated with the grading of salivary LDH5 activity in this study, predicting the aggressiveness of different histological grandes OSCC based on the reported grading system. The data suggested that LDH5 could be considered as a diagnostic parameter that should be included for detection of OSCC. The study in a large OSCC patient cohort also showed markedly increased expression of LDH5 [204]. Additionally, the elevated expression of LDH5 was found to be associated with HIF-1α and may contribute to tumor meatastatis and formation. This further supported the idea that LDH5 is a potential predictive marker of OSCC.

Lactate Dehydrogenase as a Therapeutic Target

LDH has been proposed to be a potential therapeutic target for treatment of cancers as well as other diseases. Attempts have been made to identify compounds and antibodies that effectively inhibit LDH. Cetuximab is a monoclonal antibody approved for treatment of head and neck cancer, and colorectal cancer in combination with chemotherapy/radiotherapy [205, 206]. Cetuximab binds to EGFR and inhibits the ligand-induced activation of EGFR downstream signaling, arresting cell cycle at the G1 phase which leading to apoptosis [207]. It has been shown that cetuximab-induced down-regulation of HIF-α leads to inhibition of LDHA. Interestingly, suppression of LDHA resulted in improved response of cancer cells to cetuximab [208]. This indicated that cancer cells acquire energy for proliferation from aerobic glycolysis (Warburg effect) by directing pyruvate toward the formation of lactate, the reaction catalyzed by LDH. Knockdown of LDHA expression or inhibition of LDHA activity caused a reduction in cell proliferation in cancer cells. This study suggested that LDHA is a potential therapeutic target for predicting cetuximab response in cancer cells.

Inhibition of LDHA by a drug-like small molecule, FX11 ([3-dihydroxy-6-methyl-7-(phenylmethyl)-4- propylnaphthalene-1-carboxylic acid]), revealed the mechanism underlying the tumorigenesis in lymphoma B cells [209]. It was found that LDHA plays an important role in tumor initiation as well as maintenance and progression. Suppression of LDHA expression by siRNA increased oxygen consumption in human Panc (P) 493 B-lymphoid cells, leading to remarkable ROS production and significantly elevated cell death. Inhibition of LDHA by FX11, an inhibitor specific to LDHA, also caused an increase in oxygen consumption, ROS production and necrosis. FX11 was found to inhibit cell proliferation via inhibition of glycolysis and impairment of mitochondrial respiration, resulting in decreased ATP production. FX11 also caused significant reduction of cell growth and inhibition of tumorigenesis which indicates that LDHA is essential for tumor initiation. Furthermore, inhibition of LDHA by FX11 reduced tumor xenograft progression, supporting the notion that LDHA is important for tumor maintenance.

LDHA was also suggested to be a potential therapeutic target for GC. Inhibition of LDHA by oxamate, a competitive inhibitor of pyruvate binding to the enzyme active site, caused a reduction in glycolytic metabolism in which decreased lactate formation was observed. This led to an alteration in cell morphology, reduction in proliferation and invasiveness of GC cells. Additionally, suppression of LDH also led to early apoptosis of cancer cells [210]. Supporting evidence was provided to confirm the inhibition of LDHA by oxamate in GC [187]. Oxamate inhibited cell viability in a time- and dose-dependent manner, inducing protective autophagy in GC cells. Furthermore, inhibition of LDHA by oxamate resulted in elevated ROS production.

A new competitive inhibitor of NADH binding to LDH, galloflavin (GF), was reported [211]. GF has been shown to disrupt aerobic glycolytic pathway and reduce cell proliferation in various cancer cells, including breast, colon, liver and endometrial cancer [212-215]. In endometrial cancer cells, GF effectively decreased LDHA activity, reducing cell viability, migration and colony formation in a dose-dependent manner [215]. Inhibition of LDHA activity by GF resulted in reduced rate of glucose uptake and ATP production, leading to increased expression of PDH and formation of pyruvate. It was also demonstrated that GF enhanced cell stress and apoptosis in endometrial cancer cells. This supported the notion that LDH is a potential therapeutic for cancer treatment.

Recently, derivatives of pyrrol-2-yl methanone (or ethanone), PM1 and PM2, were synthesized and their inhibitory effect against LDHA was examined by enzyme and cell-based assays [216]. *In vitro* results indicated that PM1 and PM2 altered the production of lactate as well as inhibited cell proliferation in a hepatocarcinoma cell line (SMMC 7721).

Because LDH is overepxressed in many types of cancer, it is a potential therapeutic target for cancer treatment. Figure 21 shows the inhibitors and antibody that exhibit inhibory effect on LDH.

LDH was also proposed as a biomarker for other disease conditions such as cardiovascular disease and encephalopathy. Circulating levels of LDH were increased in black South Africans who are prone to cardiovascular disease that may be related to extracellular matrix remodeling and oxidative stress [217]. It was demonstrated that rapidly growing pulmonary microvascular endothelial cells expressed a greater level of LDHA when compared with slow-growing pulmonary artery endothelial cells. Increased glucose metabolism and lactic acidosis were also observed in pulmonary microvascular endothelial cells, suggesting that cell proliferation acquires energy from aerobic glycolytic metabolism [218]. Additionally, inhibition of LDHA reduced cell growth. On the contrary, evidence has suggested the inverse correlation between normal-range LDH and development of cardiovascular disease. The association between normal-range LDH levels and risk of cardiovascular disease was investigated in a cross-sectional study of 2,087 adults [219]. Individuals who engage in more physical activity on a daily basis show higher normal-range LDH levels. Increased normal-range LDH levels were associated with lower risk of atherosclerotic cardiovascular disease, providing diagnosis value for assessment of atherosclerotic cardiovascular disease risk. However, more evidence is necessary to understand the association between LHD level and the incidence of cardiovascular disease.

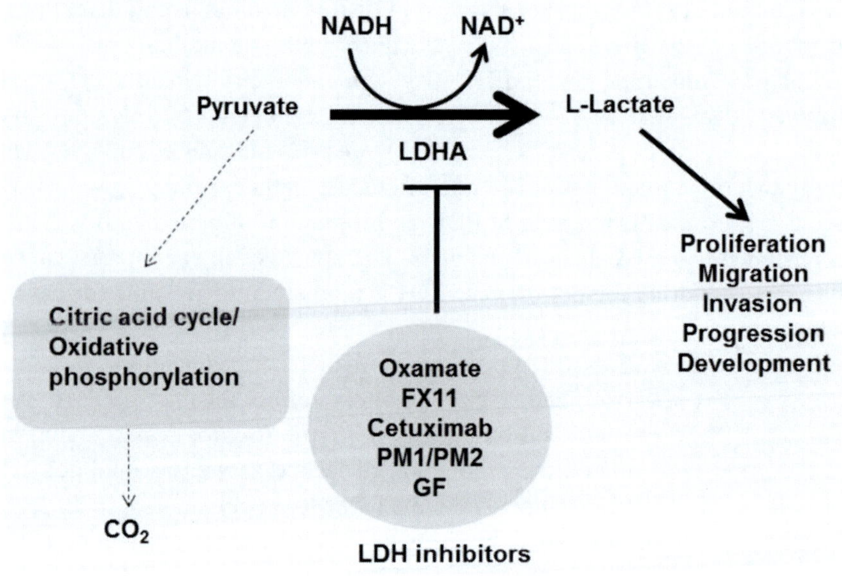

Figure 21. Inhibition of LDH. Because LDH is crucial for tumorigenesis and is highly expressed in many cancers, it is a therapeutic target for cancer treatment. Attempts have been made to indentify inhibors and antibody that effectively inhibit LDH. Oxamate, FX11, PM1/PM2 and GF are specific inhibitors of LDHA. Cetuximab is a monoclonal antibody reported to effectively inhibit LDHA.

Serum LDH concentration was found to be significantly correlated with the score of brain edema distribution in posterior reversible encephalopathy syndrome patients [220]. Approximately 85.7% of patients showed an increased serum LDH concentration of more than 226 U/L. An increased level of serum LDH was also observed in brain edema when compared with the normal level [221]. Additionally, brain lesions were associated with higher levels of LDH (792.40±186.08 vs 605.48±184.51 U/L) [222]. This indicated that LDH is an indicator of endothelial damage and increased LDH levels may function as a predictive marker of posterior reversible encephalopathy syndrome.

Moreover, the clinical role of LDH in sickle cell anemia was also studied. It was believed that serum levels of LDH could serve as a marker of hemolysis and a predictive risk marker of morbidity and mortality in sickle cell anemia [223]. A cross-sectional study in sickle cell anemia was conducted in Congolese patients [224]. Sickle cell anemia patients exhibited higher levels of serum LDH than non-sickle cell anemia subjects. It was proposed that increased serum LDH resulted from oxidative stress associated with chronic hemolysis. Additionally, elevated LDH was correlated with decreased hemoglobin rate and red blood cells, with increased white blood cells, reticulocytes and platelet count, suggesting the clinical utility of LDH as an indicator of sickle cell anemia severity.

ALCOHOL DEHYDROGENASE

Alcohol dehydrogenase (ADH; EC 1.1.1.1) is a cytosolic enzyme catalyzing the reversible oxidation of alcohols to aldehydes together with the formation of NADH, Figure 22. It is responsible for the detoxification of ingested alcohol to acetaldehyde, a highly toxic substance, which is then converted to less toxic compound, acetate, by aldehyde dehydrogenase (ALDH).

The human *ADH* gene family is located on chromosome 4q, spanning 367 kb and can be divided into 5 classes (ADH1–ADH5), according to the structure and kinetic characteristic of the enzyme. Human ADH enzymes exist in a dimeric form and contain Zn^{2+}, with a subunit molecular weight of 40 kDa, and are encoded by at least seven genes. Regarding the new nomenclature system, the polypeptide subunits, α, β1, β2, β3, γ1, γ2, π, χ, and μ/σ, of human ADH genes are designated *ADH1A*, *ADH1B*1*, *ADH1B*2*,

*ADH1B*3*, *ADH1C*1*, *ADH1C*2*, *ADH2*, *ADH3,* and *ADH4*, respectively (Table 1). The ADH enzyme family displays a wide range of substrate specificities, aliphatic/aromatic alcohols and corresponding carbonyl compounds, retinoids, steroids, amines, and fatty acids. ADH enzymes are localized in different tissues—Classes 1, 2, and 5 enzymes are mainly distributed in the liver, lung, and stomach, Class 3 enzymes are expressed in most tissues, and Class 4 enzymes are localized in the stomach, esophagus, and other mucosas.

ADH1 is the most important ADH isoform and is responsible for the detoxification of ethanol to acetaldehyde in the liver because of its high catalytic activity and expression compared with the other isoforms. Class I ADH is composed of α, β, and γ subunits, and is encoded by *ADH1A*, *ADH1B*, and *ADH1C* genes. ADH1 isoenzymes have different substrate preferences. Single nucleotide polymorphisms (SNPs) have been described for *ADH1B* and *ADH1C* genes in which each SNP showed differences in kinetic properties, leading to distinct alcohol metabolism in individuals [231]. The prevalence of *ADH1* variants varies among racial populations (Table 2). The *ADH1B**1 variant is present in nearly all populations. *ADH1B*2* has a high frequency in Asian populations but is relatively rare in Caucasians. The incidence of *ADH1B*3* is extremely low in Asians and Caucasians whereas it is common in African Americans. The prevalence of *ADH1C*2* is high in Caucasians but varies among Asian populations [232]. An additional *ADH1C* variant, Pro351Thr, has been described in Native American populations with a frequency of 25% [233]. Screening for this polymorphism in 34 populations around the world indicated that it is unique to Native Americans. Variants with higher activity develop an accumulation of acetaldehyde that can lead to adverse effects such as facial flushing, nausea, tachycardia, and finally to hepatotoxicity and tissue damage [231, 235].

ADH5 is involved in the detoxification of exogenous formaldehyde as well as in the protection of cells against toxicity by endogenous formaldehyde [236].

ADH7 was first discovered in the stomach but is highly expressed in epithelial tissues of the upper digestive tract such as the esophagus, gingival tissues, and tongue [237, 238]. This enzyme is responsible for the detoxification of alcohol in the upper digestive tract and UV-induced peroxidic aldehyde in cornea, as well as retinoic acid synthesis in epithelium tissues. It has been proposed that ADH7 might have an important role in the pathogenesis of alcohol-induced oral/esophageal cancer, which is a consequence of heavy drinking.

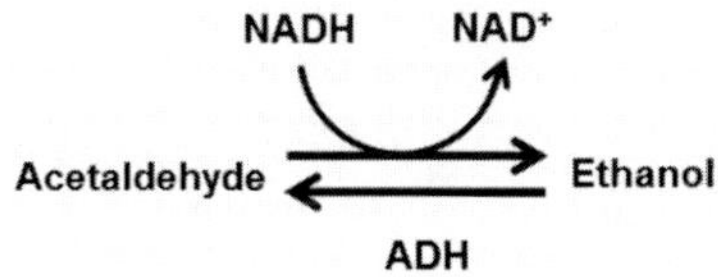

Figure 22. Reaction catalyzed by ADH.

Table 1. *ADH* genes and their properties

Class	Gene	Subunit	K_M (EtOH) mM	V_{max} (/min)	Localization	Reference
I	*ADH1A*	α	4	54	Liver	[225, 226]
I	*ADH1B*1*	β1	0.05	9	Liver, lung	[225, 226]
I	*ADH1B*2*	β2	0.9	400	Liver, lung	[225, 226]
I	*ADH1B*3*	β3	34	300	Liver, lung	[225, 226]
I	*ADH1C*1*	γ1	1.0	87	Liver, stomach	[225, 226]
I	*ADH1C*2*	γ2	0.63	35	Liver, stomach	[225, 226]
II	*ADH4*	π	34	40	Liver, cornea	[227]
III	*ADH5*	χ	1,000	-	Most tissues	[228]
IV	*ADH7*	μ/σ	20	1,510	Stomach, oesophagus and other mucosas	[229, 230]

Table 2. Polymorphism of Class I *ADH*

Gene	Mutation	Population with high allele frequency	Reference
*ADH1B*1*	Wild-type: Arg 48, Arg370	Caucasians, American-Africans	[234]
*ADH1B*2*	Arg48His allele	Asians	[234]
*ADH1B*3*	Arg370Cys allele	American-Africans	[234]
*ADH1C*1*	Wild-type:Arg272, Ile350	All citizens	[234]
*ADH1C*2*	Arg272Gln/Ile350Val	Caucasians	[234]
*ADH1C*3*	Arg272Gln/Ile350Val	Native Americans	[233]

Alcohol Dehydrogenase, Alcohol and Liver Disease

ADH is mainly responsible for the detoxification of ethanol and *ADH1* polymorphisms have been shown to be an important genetic factor in alcohol metabolism and alcoholism. Indeed, several determinants are conducive to alcoholism such as duration, exposure time, concentration of ethanol, and activity of alcohol-metabolizing enzymes. The activity of two important alcohol-metabolizing enzymes, ADH and ALDH, was shown to be associated with the level of acetaldehyde. Increased activity of ADH and decreased activity of ALDH can result in an accumulation of acetaldehyde, causing several adverse effects such as headache, nausea, and high heart rate. Furthermore, the activity of ADH is correlated with polymorphisms within its gene. As a consequence, the effects of *ADH1* variants on susceptibility to alcoholism have been evaluated [239]. Meta-analysis studies reported that *ADH1B*1* is associated with an increased risk of alcoholism although it exhibits a protective effect against liver diseases in East Asian populations [240, 241].

Many studies have indicated that *ADH1B*2* contributes to a reduced risk of the development of alcohol dependence while increasing the risk for the development of alcoholic liver disease among alcoholics, in particular, in Asians [240, 242-244]. Furthermore, *ADH1B*2* was reported to promote the alcohol elimination rate in Jewish populations when compared with the wild-type gene [245]. A further study was performed to investigate the effects of *ADH1B*2* in Japanese populations [246]. In agreement with the previous study in Jews, an enhanced alcohol elimination rate was observed in Japanese populations carrying the *ADH1B*2* genotype. A meta-analysis study provided strong evidence of the involvement of *ADH1B*2* in the pathogenesis of alcohol dependence, including alcohol-induced diseases such as cirrhosis, hepatitis, and alcohol-induced pancreatitis [243]. Screening a large number of SNPs in all seven *ADH* genes indicated that not only *ADH1B* is associated with the susceptibility to alcoholism; *ADH1A*, *ADH2*, *ADH3,* and *ADH4* genes are also involved [247, 248].

A genetic association and gene-gene interaction study in Indian populations revealed that genetic variations in *ADH1B*3* were associated with an increased risk of alcoholic liver disease [249]. Because *ADH1B*3* is mainly found in African American populations, it was suspected that they have an increased risk for adverse outcomes after intrauterine ethanol exposure. However, the *ADH1B*3* polymorphism was associated with protection from adverse outcomes, birth weight, and mental development in infants at 1 year of age [250]. This is because this variant encodes an enzyme with high catalytic activity, enhancing alcohol elimination in carriers. Although the protective effect of *ADH1B*3* is significant for both offspring growth and development, the magnitude of this effect is relatively small. Therefore, additional factors must be taken into account for the varying susceptibility of African American infants exposed to alcohol.

The effect of the *ADH1C* polymorphism on alcoholism and liver cirrhosis has been widely investigated. A significant association was found between *ADH1C*1* and alcoholic liver cirrhosis whereas there was no correlation for *ADH1B* variants [251]. Interestingly, the risk of developing liver cirrhosis was extremely high in patients carrying a combination of *ADH1C*1* and *ADH1B**2 or *CYP2E1*5B* (cytochrome P450 2E1) or a null genotype of *GSTM1*, suggesting cross talk between enzymes involved in alcohol metabolism and radical detoxification. Because the *ADH1C*1* genotype encodes a more highly active enzyme (2.5-fold higher activity) when compared with *ADH1C*2*, it was speculated that levels of aldehyde and ROS are increased that could lead to lipid peroxidation and further generate radical species causing liver damage.

It was revealed that *ADH1C*2* has a significant effect on alcohol dependence in Asian subjects but not in European populations [240, 242]. However, *ADH1C*2* had no effect on alcoholism or liver diseases in European populations [234]. The combined effect of *ADH1B*3* and *ADH1C*2* has been described in Hungarian populations and it was found that *ADH1C*2* contributed to alcohol drinking behavior while a combination of *ADH1B*2* and *ADH1C*1* conferred a protective effect against chronic liver disease [252]. Significantly high frequencies of fast alcohol metabolizing variants (*ADH1B*2* and *ADH1C*1*) were found in healthy individuals compared with alcoholics in the Taiwanese population [253]. The frequencies of *ADH1B*2* and *ADH1C*1* genotype and *ADH1B*1* and *ADH1C*1* genotype were significantly different between drinkers and nondrinkers. Additionally, the prevalence of *ADH1C*1* and *ADH1C*2* was not different between controls and alcoholics. Because Class I *ADH* genes are located within 80-kb of chromosome 4, it was hypothesized that both variants could not be inherited independently. Therefore, it was believed that linkage disequilibrium occurred at functional *ADH1C*1* and at *ADH1B* and contributed to alcohol dependence in individuals carrying these genotypes. Linkage disequilibrium was also proposed for *ADH1B*2* and *ADH7*, conferring a protective effect against alcoholism [254]. Furthermore, the presence of the *ADH1B*2* polymorphism alone might not be sufficient to account for protection against alcohol dependence, and an additional allele might be required to mediate this effect.

Alcohol Dehydrogenase and Brain Damage

The association between the *ADH1B*1* variant and alcoholic brain atrophy has been reported [255]. This is because the rate of alcohol oxidation in *ADH1B*1* carriers is slower than that of *ADH1B*2*, causing an accumulation of ethanol, which causes brain damage.

ADH is thought to contribute to the pathogenesis of neurodegenerative disorders because of its important role in alcohol, retinoid, and dopamine metabolism in many cell types, including neurons. An association between *ADH4* variants and PD was found in Swedish populations but was absent in Americans [256]. Although it was suggested that a polymorphism in the *ADH4* gene might be a risk factor for the development of PD, *ADH4* variations alone are not sufficient to cause the disease. Therefore, it was suggested that other genetic and/or environmental factors might also contribute to PD. A SNP in *ADH1C* (rs283413), causing a stop codon at residue 78 in the protein, was significantly associated with PD in European ancestry and Caucasian patients [257]. The importance of ADH1 and ADH4 in PD was examined in a mice model where $ADH1^{-/-}$, $ADH4^{-/-}$, and $ADH1/4^{-/-}$ mice had a reduced sense of smell and altered dopamine metabolism, which is in agreement with the reported olfactory dysfunction in PD patients. Additionally, changes in spontaneous locomotor activity were observed in these mice, indicating they might be useful for studying presymptomatic PD [258, 259].

Alcohol Dehydrogenase and Cardiovascular Disease

In addition to alcoholism, the *ADH1C* polymorphism has been hypothesized to play a role in cardiovascular disease. A decreased risk of myocardial infarction was associated with the *ADH1C*2* polymorphism, which encodes a slower ethanol oxidizing enzyme in comparison to the *ADH1C*1* polymorphism [260]. Individuals with moderate alcohol consumption are at reduced risk of myocardial infarction independent of the *ADH1C* genotype. However, the *ADH1C*2* variant showed the greatest reduction for the development of myocardial infarction, which was also associated with plasma high-density lipoproteins (HDL) levels. Furthermore, a study in Massachusetts showed that the *ADH1C*2* polymorphism reduced the risk of cardiovascular disease where carriers of the rs1693482 homozygous had a stronger association between alcohol and cardiovascular disease than those with rs698 homozygous [261]. However, the association between the

ADH1C genotype and HDL cholesterol was not found in this study. *ADH1C*2* was found to be a cardioprotective genotype in Caucasian British men (modest alcohol consumption) and also confers protection against coronary heart disease [262].

Alcohol Dehydrogenase and Cancer

Alcohol consumption is considered as one of the factors contributing to ESCC caused by the exposure of tissues to alcohol or its metabolite, acetaldehyde. Therefore, the effects of polymorphisms in alcohol metabolizing enzymes such as ADH and ALDH have been investigated. The susceptibility of those with the *ADH1B* polymorphism to the development of ESCC was reported in the Japanese, Taiwanese, and European populations [263-265]. A 3.99-fold increased risk of developing esophageal cancer was found for the *ADH1B*1* variant compared with the *ADH1B*2* polymorphism in Taiwanese populations [263]. The *ADH1B*1* variant was more prevalent in alcoholic subjects than healthy persons in a Japanese population and a significant association was found between the *ADH1B*1* polymorphism and an increased risk of developing ESCC [264]. Additionally, the effect was more pronounced when *ADH1B*1* was present in combination with the *ALDH2* variant or heavy alcohol drinking. Interestingly, a study of Turkmens and non-Turkmens in Iran revealed a decreased risk to ESCC by the *ADH1B*2* polymorphism under a recessive model [265]. Compared with the *ADH1B*1* variant, carriers of two *ADH1B*2* alleles were at reduced risk of ESCC.

A genome-wide association study (GWAS) in Chinese populations has revealed new ESCC susceptibility loci in *ADH* genes (rs1042026 and rs17033 in *ADH1B*, rs1614972 and rs1789903 in *ADH1C* and rs17028973 in *ADH7*), which are significantly associated with alcohol drinking behavior in relation to the increased risk of developing ESCC [266]. Carriers of fast ADH oxidizing enzymes are expected to have a higher risk of this cancer. The study demonstrated a joint effect of *ADH1B* and *ALDH2* polymorphisms and alcohol intake on ESCC in which a combination of *ADH1B* and *ALDH2* alleles has the highest risk of ESCC development in alcohol drinkers. Another GWAS analysis was also demonstrated the susceptibility of these polymorphisms (rs1042026, rs17033, rs1614972, rs1789903 and rs17028973) in a cluster of *ADH1B- ADH1C- ADH7* for the progression of esophageal cancer in Chinese alcoholics [267]. Subjects who are heavy alcohol drinkers with the susceptible

*ADH1B*1* genotype are at a higher risk for developing ESCC than those who are nondrinkers and do not carry the *ADH1B*1* genotype.

A population-based case control study in China also demonstrated the association of the *ADH1B*1* genotype and an increased risk esophageal cancer independent of alcohol consumption and tobacco smoking behaviors [268]. Additionally, an interaction between *ADH1B*1* and *ALDH2* was observed and contributed to esophageal cancer susceptibility. However, the association between *ADH1C*2* and ESCC was not reported in this study. The suggested mechanism for the increased risk in *ADH1B*1* carriers to ESCC might be explained by the longer exposure to acetaldehyde in the circulation of individuals who are slow alcohol metabolizers. During high exposure to acetaldehyde in *ADH1B*1* carriers, DNA binds to acetaldehyde causing mutations in procarcinogenic or apoptotic genes that might cause cancer. A 2-fold increased risk of developing of esophageal and gastric cardia adenocarcinoma was found in American populations carrying the *ADH1C*1* genotype [269]. Taken together, it was demonstrated that *ADH* genes play an important role in the etiology of ESCC and genetic polymorphisms of *ADH* are critical factors for cancer development. However, cancer is a multifactorial process; therefore, other factors such as genetic mutations and environmental changes should also be taken into account.

In addition to its protective effect against alcoholism, *ADH1B*2* was reported to confer protection against esophageal and upper respiratory cancers. The proposed protective effect was based on a hormesis mechanism in which a limited (short) exposure to acetaldehyde might enhance DNA repair in *ADH1B*2* (a slow alcohol metabolizing variant) drinkers [270]. This effect is in contrast to the fast oxidizing enzyme *ADH1B*2* variant, which leads to a longer exposure of DNA to acetaldehyde and potential DNA damage. The evidence for hormesis was also observed for low doses of oxidizing radiation to DNA repair and for low acetaldehyde concentrations to increased longevity in *Drosophila* [271, 272].

Evidence from epidemiological analyses suggested an association between alcohol consumption and breast cancer [273, 274] based on exposure to acetaldehyde. Indeed, both *in vivo* and *in vitro* studies have indicated a carcinogenic role for acetaldehyde [275, 276]. A population-based case-control study in the German population indicated the protective role of *ADH1B*2* against breast cancer. A reduced chance of developing breast cancer was found in heavy alcohol drinkers carrying the *ADH1B**2 genotype, suggesting that faster alcohol elimination might reduce the risk for breast cancer.

The *ADH1C* genotype was found to modify the association between alcohol consumption and breast cancer [277]. A case-control study of Caucasians indicated an increased risk of developing breast in premenopausal women carrying the *ADH1C*1* polymorphism with high alcohol consumption. Likewise, a population-based study of Caucasian women in Long Island indicated an increased risk of breast cancer among alcohol drinkers who carry the *ADH1C*1* polymorphism, a faster metabolizing enzyme compared with the *ADH1C*2* genotype [278]. An alternative explanation might be the effect of alcohol on the levels of steroid hormone, which might have an important role in breast carcinogenesis [279, 280]. However, this is still inconclusive because many other studies did not find a significant correlation between alcohol drinking and endogenous hormone levels [281-283]. A prospective study of the *ADH1C* polymorphism showed no relationship between *ADH1C* variants or alcohol intake and breast cancer risk in American populations [283]. However, the *ADH1C* polymorphism moderately affected the response of steroid hormones to alcohol intake, which was not associated with breast cancer risk.

Interactions between alcohol consumption and tobacco smoking are believed to contribute to HNSCC. A role for *ADH1C* polymorphisms in HNSCC susceptibility was reported in Massachusetts citizens [284]. Alcohol consumption and tobacco smoking resulted in an elevated risk of HNSCC. The *ADH1C*2* genotype conferred a significant increase for developing HNSCC in heavy drinkers. Nonetheless, a less dramatic effect of the *ADH1C*2* polymorphism on HNSCC susceptibility was observed in heavy smokers. In contrast, a hospital based case-control study in Iowa State showed no correlation between the *ADH1C* polymorphism and the risk of developing HNSCC [285]. The *ADH1C* polymorphism did not cause an alteration in the association of alcohol consumption and smoking and HNSCC. However, they found a poorer survival for carriers of the *ADH1C*1/2* genotype when compared with other *ADH1C* polymorphisms. Similarly, no association was found between the *ADH1C**1 polymorphism and a risk of developing HNSCC in a hospital-based study [286] and a meta-analysis study [287]. *ADH1C**1 and *ADH1B**2 polymorphisms conferred a protective effect against the progression of HNSCC in a meta-analysis study [288]. The effect was stronger in Asian populations than Caucasians possibly because *ADH1C**1 and *ADH1B**2 polymorphisms are more prevalent in Asian countries.

However, it was demonstrated that the *ADH1B**2 polymorphism is significantly associated with an increased risk of HNSCC in Asian populations [289]. This effect was also reported in the Korean population where the

*ADH1B**2 polymorphism and *ADH1C* Iso350Val are increased risk factors for HNSCC [290].

It was proposed that alcohol consumption might contribute to lung cancer. An association between alcohol intake and increased risk of lung cancer was reported in a hospital-based case control study [291]. However, there was no clear evidence of the involvement of the *ADH1C* polymorphism in lung cancer in this study. They found an association of *ALDH2* variants with increased risk of lung cancer. ADH1B expression was decreased in lung cancer, indicating an association with this type of cancer [292]. Because ADH1B is involved in several metabolic pathways including fatty acid, retinol, tyrosine, glycolysis, and gluconeogenesis, it is believed to play a part in lung cancer. However, no correlation between the *ADH1C* polymorphism and lung cancer was found in a case-control study in Erie and Niagara countries [293].

Chronic alcohol consumption is a risk factor for colorectal cancer. It was demonstrated that polymorphisms in alcohol metabolizing enzymes cause an accumulation of acetaldehyde that might be responsible for the carcinogenic effect [275, 294]. The total activity of ADH1 in the serum was significantly higher in colorectal cancer patients than in healthy persons [295]. Furthermore, an increased activity of ADH1 was present in accordance with the advanced status of the disease. This suggested that alterations in ADH1 activity are caused by the isoenzyme release from cancer cells. A study in a Dutch population indicated a relationship between alcohol consumption and the increased risk of colorectal adenoma [296]. Additionally, it was demonstrated that the *ADH1C*1* genotype plays a role in the association between alcohol intake and cancer because a significantly increased risk was found in high alcohol consumption subjects carrying *ADH1C*1*. It was also suggested that aldehyde is a carcinogen. Similarly, *ADH1C*1* was associated with the increased risk of colorectal cancer in heavy drinkers (more than 30 g of ethanol per day) in Japanese populations [297]. An association between the *ADH1B*2* polymorphism and colorectal cancer was not observed in Caucasian populations [244]. However, the *ADH7* variant (rs1573496(C)) might alter the relationship between alcohol consumption and colorectal cancer.

Many studies have reported a significant association between the *ADH1C*1* polymorphism and cancer risk [278, 298]. This might be because of the higher catalytic activity of *ADH1C*1* than *ADH1C*2*, producing higher levels of acetaldehyde, which is a key player in carcinogenesis. A meta-analysis study (35 case-control) of *ADH1C* reported that *ADH1C* Ile350Val was not associated with the risk of overall cancer. However, an increased cancer risk was found for *ADH1C* Ile350Val among African and Asian populations but not Europeans or those of mixed race [299].

Serum levels of *ADH1C,* possibly released by cancer cells, were higher in pancreatic cancer patients. Therefore, *ADH1C* might be a candidate marker for pancreatic cancer [300]. The association between increased cancer risk and the *ADH1C*1*genotype was evident in heavy drinkers [298]. Furthermore, high alcohol consumption Caucasian patients homozygous for *ADH1C*1* have a greater risk of developing cancers of the oral cavity (oropharynx, hypopharynx and esophagus) and liver, suggesting *ADH1C*1* might be a genetic marker for cancer. This is in contrast to a study in America in which no association was found between *ADH1C* variants and oral cavity and pharynx carcinomas [301]. The association between *ADH1C* genotypes and the risk of developing cancer is therefore controversial. This might be explained by small study populations obscuring the effect of the *ADH1C* genotype or the lack of specific alcohol intake information such as drinks/week. An alternative explanation might be the different ethnicity between different populations in each study.

Alcohol Dehydrogenase and Pancreas/Liver Diseases

Many studies have investigated the association between the *ADH1C* polymorphism and pancreatitis. A study in an Indian population indicated the susceptibility of *ADH1C* variants to alcoholic pancreatitis [302]. A strong association between the *ADH1C*1* genotype and alcoholic pancreatitis was observed. A link between the *ADH1C* polymorphism and alcoholic pancreatitis was also reported in other ethnic groups such as English patients. However, in other racial populations (Caucasians, Australian and Dutch), no correlation between *ADH1C* variants and pancreatitis was observed [303, 304]. Therefore, it was proposed that the susceptibility of those with the *ADH1C* polymorphism to alcoholic pancreatitis depends on the ethnic group. Nonetheless, a recent meta-analysis study (nine eligible studies included) suggested that the *ADH1C*1* genotype contributes to pancreatitis, especially chronic alcoholic pancreatitis [305].

Although no relationship between the *ADH1C*2* polymorphism and alcoholic pancreatitis was reported, it was revealed that it is associated with liver cirrhosis in Australians, in contrast to other reports [303]. One possible explanation for this discrepancy is that other environmental factors might be involved and therefore contribute to the pathogenesis of liver cirrhosis.

Additionally, the *ADH1C*2* genotype might be associated with alcohol dependence and increase the susceptibility to cirrhosis.

Taken together, contribution of ADH to disease comes from polymoriphisms in the gene, resulting in different enzyme activities. Acetaldehyde is a toxic substance, when it is accumulated in cells it can be harmful, causing liver and DNA damages. Acetaldehyde can cause mutations in procarcinogenic and apoptotic genes which can lead to development of cancer. The role of ADH in disease is summarized in Figure 23.

The role of the *ADH3* polymorphism in the development of steatohepatitis has been investigated. It was demonstrated that ADH3 confers a protective effect against steatohepatitis in an NRF2-independent manner. Additionally, ADH3 prevented the development of steatosis to steatohepatitis by maintaining the level of cellular GSH to provide a sufficient amount of antioxidant through an NRF2-mediated reaction [306].

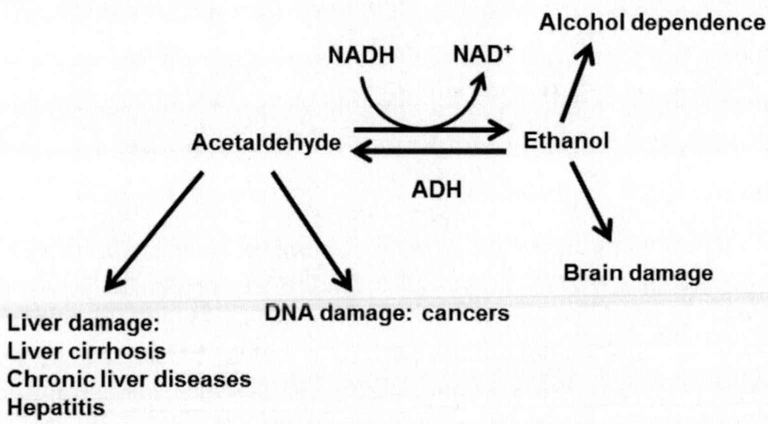

Figure 23. The role of ADH in development of diseases. Accumulation of acetaldehyde is toxic to cells, resulting in damage to liver and DNA. Acetaldehyde is a carcinogen, causing mutations in procarcinogenic or apoptotic genes that might lead to cancer. Accumulation of ethanol also causes brain damages.

ISOCITRATE DEHYDROGENASE

Isocitrate dehydrogenase (IDH; EC 1.1.1.42) is an enzyme that catalyzes the reversible oxidative decarboxylation of isocitrate to α-ketoglutarate (α-

KG) and carbon dioxide together with the formation of NADH or NADPH in which divalent ion cofactors (Mg^{2+} or Mn^{2+}) are also required for catalysis (Figure 24). The IDH protein family consists of 3 members—IDH1, IDH2 and IDH3—which have different preferences for coenzymes, including their localization, Table 3 and Figure 25.

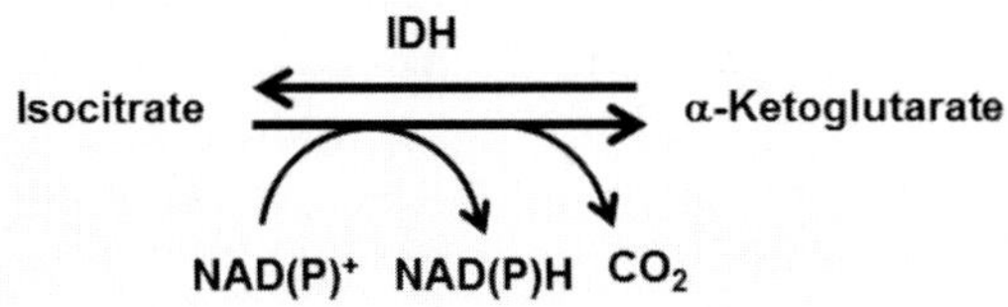

Figure 24. Reaction catalyzed by IDH.

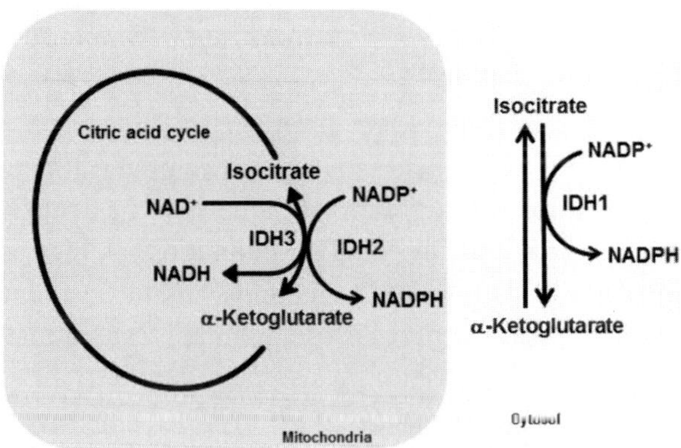

Figure 25. Compartmentation of reactions catalyzed by IDH enzymes. Reaction catalyzed by IDH occurs in cytosol while IDH2 and IDH3 are involved in transformation of isocitrate to α-KG in citric acid cycle.

Table 3. IDH enzymes and their properties

Isoform	Coenzyme	Localization	Active form
IDH1	$NADP^+$/NADPH	Cytosol, peroxisome	Dimer
IDH2	$NADP^+$/NADPH	Mitochondria	Dimer
IDH3	NAD^+/NADH	Mitochondria	Tetramer

IDH1 and IDH2 are homodimeric enzymes with approximately 69% amino acid identity, sharing an almost identical 3-dimensional structure. However, IDH3 is heterotetrameric and does not have sequence similarity with

IDH1 and IDH2. IDH3 catalyzes a critical step in the citric acid cycle of mitochondria. Somatic mutations in the genes encoding IDH1/2 are often observed in human tumors. The most common mutation in IDH1 is found at arginine 132, which can be changed to histidine, serine, cysteine, glycine, or leucine, where Arg132His was the most frequent mutant [307]. Other mutations in IDH1 were also reported, Val71, Glu97, Arg100, Gly123 and Tyr139 [308-310]. In IDH2, Arg172, which is analogous to IDH1 Arg132, was exclusively present in glioma [307] and was mutated to lysine, glycine, serine, methionine or tryptophan [311]. The mutation at Arg140, which is equivalent to IDH1 Arg100, frequently present in acute myeloid leukemia (AML) [310]. A mutation to glutamine and leucine was found at residue 140 in IDH2 [311].

Isocitrate Dehydrogenase and Cancer

The first mutations in IDH were discovered in metastatic colon cancer [312]. IDH1/2 mutations were later found in low-grade gliomas and also a subset of secondary glioblastoma with a frequency of 50–80%. However, a much lower frequency (5%) of the IDH1/2 mutation was reported in primary glioblastoma [307, 313, 314]. IDH1/2 mutations might be founder mutations because they occur early in the process of malignant transformation, followed by other subsequent mutations that may also be required to drive oncogenesis [314]. IDH mutations were also discovered in myeloid malignancies, including myeloproliferative neoplasms, myelodysplastic syndromes, and AML [313, 315, 316]. In 2009, whole genome sequencing identified an IDH1 Arg132His mutation in AML and subsequent studies reported the presence of IDH2 Arg140 and Arg172 mutations [315]. The reported rates of IDH1/2 mutations are in the range of 6-11%. Additionally, an IDH1 mutation, Arg132Lue was identified in breast adenocarcinoma [317]. IDH2 Arg172 mutations were also reported in angioimmunoblastic T-cell lymphoma (AITL) [318].

The investigation of IDH1/2 mutations in thyroid cancers showed no mutation at residue 132 of IDH1 and 172 of IDH2; however, a novel mutation (Gly123Arg) was identified in IDH1 with a frequency of 11.11% [309]. IDH1/2 mutations were also observed in chondroma and cholangiocarcinoma [319-321]. Furthermore, IDH1 Arg132Cys was identified by deep sequencing in medulloblastoma [322]. The IDH1 Arg132Cys mutation was also found in inflammatory bowel disease–associated intestinal adenocarcinoma but not in intestinal adenocarcinoma [323] and was exclusively present in microsatellite stable colorectal cancers [324]. An IDH2 Arg172Ser mutation was frequently present in osteosarcoma as studied by DNA sequencing and by a multi-specific monoclonal antibody, MsMab-1 [325]. IDH1/2 mutations found in different types of cancer are shown in Table 4.

Table 4. IDH mutations identified in cancers

Cancer	IDH1 mutation	IDH2 mutation	References
AML, myeloid dysplastic syndrome, myeloid neoplasms	Arg132His Arg132Cys	Arg140Gln Arg140Trp	[308, 315, 316, 326]
	Arg132Ser	Arg140Leu	
	Arg132Gly	Arg172Lys	
	Arg132Leu	Arg172Trp	
	Val71Ile		
	Ile99Met		
AITL	-	Arg140Gly	[318, 327]
	-	Arg172Lys	
	-	Arg172Gly	
	-	Arg172Thr	
	-	Arg172Ser	
Breast adinocarcinoma	Arg132Lue	-	[317]
Colorectal cancers	Arg132Cys	-	[324]
Cholangiocarcinoma	Arg132Cys	Arg172Trp	[320]
	Arg132Leu	-	
	Arg132Gly	-	
Chondroma, Chondrosarcoma, Osteosarcoma	Arg132Cys	Arg172Ser	[319]
	Arg132Ser	Arg172Thr	
	Arg132Gly	-	
	Arg132Leu	-	
	Arg132His	-	
Glioma	Arg132His Arg132Cys	Arg172Lys Arg172Met	[307, 311, 313, 314]
	Arg132Ser	Arg172Trp	
	Arg132Gly	Arg172Gly	
	Arg132Leu	Arg172Ser	
	Arg132Val	-	
Medulloblastoma	Arg132His	-	[322]
Osteosarcoma	-	Arg172Ser	[325]
Thyroid cancer	Gly123Arg	-	[309]
	Val71Ile	-	

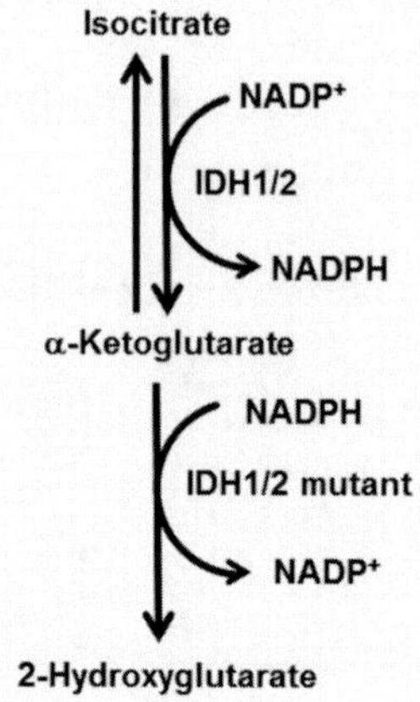

Figure 26. Reaction catalyzed by IDH mutants. α-KG is transformed into 2-hydroxyglutarate together with the conversion of NADPH to $NADP^+$.

However, IDH3 mutations have not been found in human tumors, possibly because IDH3 is a critical enzyme in the citric acid cycle, and disrupting IDH3 function might lead to significant growth retardation. Tumor cells usually carry one mutated allele and a native allele. Point mutations affect the catalytic activity of enzymes leading to a loss of the native activity and the gain of a novel enzymatic function [311]. The ratio of mutation to wild-type in these genes is believed to play important role in the oncogenic activity of the enzyme.

A point mutation, Arg132His, in IDH1 was demonstrated to be a key amino acid in the active site of the enzyme [311]. Arg132 is highly conserved in all bacterial and eukaryotic IDH1 enzymes [328]. Replacing Arg132 with His altered the active site of the enzyme, leading to an impaired binding affinity for substrate, isocitrate, and cofactor Mg^{2+}, which caused a 1,000-fold decrease in catalytic turnover. Interestingly, Arg132His causes an accumulation of 2-hydroxyglutarate (2-HG) suggesting a novel effect of the Arg132His mutant to catalyze the conversion of α-KG to 2-HG in an NADPH-dependent manner. This notion was supported by the high level of 2-HG (50- to 100-fold) observed in malignant glioma expressing IDH1/2 mutations compared with normal tissues. Therefore, 2-HG has been termed an "oncometabolite" and is believed to contribute to the formation and malignant progression of tumors. Other Arg132 mutations in humans such as Arg132Cys, Arg132Leu, and Arg132Ser share similar features with

Arg132His. The accumulation of 2-HG in the brain causes leukoencephalopathy. Additionally, 2-HG increased the ROS levels in brain tissue and increased the risk of progression to brain tumors [329-331].

Regarding IDH2, residue 172 is located in the active site of the enzyme and is analogous to residue 132 in IDH1. Likewise, IDH2 mutants (Arg172Lys, Arg140Gln) also exhibited similar characteristics to IDH1 mutants in which the mutation caused a loss in normal activity and resulted in the gain of a new activity, producing 2-HG (Figure 26) [307].

The potential mechanism by which IDH1/2 mutations contribute to pathogenesis of cancers has been investigated. α-KG is a substrate for many α-KG-dependent dioxygenases and 2-HG produced by IDH1/2 mutants is structurally similar to α-KG. Therefore, 2-HG could act as a competitive antagonist of α-KG, disrupting the function of various α-KG-dependent dioxygenases. α-KG-dependent dioxygenases play important roles in many pathways, including collagen biosynthesis, sensing of oxygen, hypoxia response, DNA repair, chromatin modification and lipid metabolism. More than 60 dioxygenases have been reported in mammalian cells, including prolyl hydroxylases, collagen prolyl hydroxylases, histone demethylases and the ten-eleven translocation (TET) family of DNA hydroxylases [332-334]. As a consequence, an IDH1/2 mutation is expected to influence various cellular processes. In fact, it was found that 2-HG altered the function of various α-KG-dependent dioxygenases with different potencies [335, 336].

Furthermore, owing to the structural similarity between 2-HG and the excitatory neurotransmitter glutamate, 2-HG could activate the glutamate receptor, N-methyl-D-aspartic acid (NMDA), in neurons, resulting in disturbance of intracellular calcium homeostasis and generation of ROS [329, 330]. Glutamateric pathway is involved in development, proliferation, survival and metastasis of tumors [337]. Accumulation of 2-HG could lead to 2-hydroxyglutaric aciduria, an inherited metabolic disorder with severe neurological dysfunction and cerebral atrophy. 2-HG was found to increase the production of free radicals which stimulates lipid peroxidation. Additionally, 2-HG also reduced the brain's ability to cope with damage caused by accumulation of ROS. Furthermore, it is evident in glioma cells that 2-HG produced by IDH1 mutants enhanced oxidative stress, apoptosis and autophagy. However, the effects varied greatly among the different cell types. Therefore, further experiments are required to confirm the contribution of IDH1 mutants to glioma cell phenotypes. Nonetheless, this suggested that generation of oxidative stress via the effect of 2-HG might contribute to the pathophysiology of brain injury.

Methylation of DNA and histone plays a critical role in the regulation of gene expression and cellular differentiation. Methyltransferases and demethylases are two major enzymes responsible for the establishment and maintenance of DNA and histone methylation. Structural analysis has revealed that IDH1/2 mutants inhibited the function of histone demethylase and the TET family of 5-methylcytosine hydroxylases in which 2-HG occupied the same space in the active site of histone demethylase as α-KG. Histone demethylases are responsible for the removal of methyl groups from methylated lysine and controlling the gene expression by modulating chromatin remodeling and transcription factor binding. Besides, TET catalyzes the conversion of 5-methylcytosine (5-mC) to 5-hydroxymethylcytosine (5-hmC). TET plays an important role in the regulation of cellular differentiation and also maintenance of embryonic stem cells by controlling expression of pluripotency genes [338, 339]. 5-Hydroxymethylcytosine itself is involved in passive demethylation of DNA by blocking DNA methyltransferase (DNMT) and inactivates DNA methylation by DNA repair mechanisms [340]. Inhibition of these enzymes by 2-HG resulted in genome wide alteration in histones and DNA methylation, which may affect the expression of oncogenes, tumor suppressors and metabolic pathways.

Meta-analysis in AML, low-grade glioma, cholangiocarcinoma and chondrosarcoma methylation data revealed cancer-specific effectors within the retinoic acid receptor activation pathway [341]. Another study has identified transcription factor EBF1 (early B-cell factor 1) as an interaction partner for TET2. Increased methylation of histone and DNA was also observed in glioma expressing an IDH1 mutant. Furthermore, expression of IDH2 Arg172Lys altered DNA methylation in the Jurkat cell line, $CD4^+$ T cells and clinical samples of AITL [318]. In humans, alteration in histone demethylation was found to be associated with development and aggressiveness of cancer [342]. Because histone demethylase is required for differentiation of lineage-specific progenitor cells into terminally differentiated cells, inhibition of this enzyme by 2-HG in IDH1/2 mutants caused impairment of cell differentiation of non-transformed cells [343]. This was also evident in a mouse model where the IDH1 Arg132His mutation was inserted into cells of myeloid lineage. The mutation caused increased numbers of early hematopoietic progenitors, splenomegaly and anemia. Furthermore, hypermethylated histone and altered DNA methylation were also observed in a similar manner to those observed in AML patients harboring IDH mutants. It was shown that inhibition of histone

3 Lysine 9 (H3K9) demethylase by 2-HG in IDH1/2 mutants was sufficient to block the differentiation of non-transformed cells.

Another study also suggested that an IDH1 mutation is sufficient to establish a hypermethylator phenotype [344]. An IDH1 mutant caused methylome remodeling which resulted in distinct Glioma CpG island methylator phenotype (G-CIMP). The presence of the mutation altered specific histone marks, activated extensive DNA hypermethylation and reorganized the methylome, indicating that an IDH1 mutation is the molecular basis of G-CIMP in glioma and provides evidence of the interplay between epigenetic and genomic changes in human tumors. In line with this observation, heterozygous expression of IDH1 Arg132His in the human colorectal cancer cell line HCT116 was found to be sufficient to cause genome-wide alterations in DNA methylation [345]. One possible mechanism to explain the G-CIMP observed in glioma is inhibition of the TET family of DNA hydroxylases by 2-HG produced in IDH1/2 mutants [335]. Additionally, hypermethylation was noted in both glioma and AML patient samples [338, 346]. IDH1 mutants were also found to correlate with promoter methylation of the DNA repair gene O^6-methylguanine-DNA methyltransferase (MGMT), suppressing its expression and indicating strong prognostic significance for overall survival [347]. Altogether, the alterations in epigenetic status (histone modifications and DNA methylation) mentioned above have the potential to change the expression of various oncogenes and tumor suppressor genes and could provide a plausible explanation for the contribution of IDH mutations to tumorigenesis.

Two other enzymes, prolyl hydroxylases and collagen-prolyl-4-hydroxylase, were also inhibited by 2-HG, resulting in activation of HIF-1α and reduced endostatin expression [335]. HIF-1α is a key component of the transcription factor that is responsible for cellular adaptation to oxygen availability and regulation of the expression of genes involved in glucose metabolism, angiogenesis, apoptosis, metastasis and other pathways critical to tumor growth such as growth factors, invasion-promoting genes and angiogenesis-promoting genes. Therefore, changes in HIF-1α levels in IDH mutants might influence gliomas and other tumors. Endostatin suppresses angiogenesis and tumor growth.

In the presence of oxygen, prolyl hydroxylase destabilizes HIF-1α via hydroxylation of a proline residue followed by Von Hippel-Lindau (VHL)-mediated proteasome degradation [348]. Accordingly, it was demonstrated that IDH1 Arg132His and IDH2 Arg172Gly can increase HIF-1α protein levels in U-87MG, HEK293T and C6 glioma cells [349, 350]. Inhibition of prolyl hydroxylase was also observed in IDH1 mutant cell lines. Additionally, increased HIF-1α levels were only observed in glioma cells harboring the Arg132His mutant but not the wild-type. In IDH1/2 mutations, HIF-1α is accumulated and translocated into the nucleus, transcriptionally activating the expression of downstream gene targets. Furthermore, a study in a mouse model indicated inhibition of prolyl hydroxylation of HIF-1α by 2-HG and upregulation of HIF-1α target gene maturation [351]. 2-HG also inhibited collagen hydroxylation, resulting in impairment of collagen protein maturation. Accumulation of immature collagen could contribute to embryonic lethality and aberration of the basement membrane that is believed to be associated with the development of glioma. Therefore, inhibition of both prolyl hydroxylases and collagen-prolyl-4-hydroxylase by 2-HG might activate tumor growth and angiogenesis. Mechanisms by which IDH mutations contribute to tumorigenesis are depicted in Figure 27.

IDH1 mutations were also found to cause aberration in glucose metabolism. Downregulation of the citric acid cycle was observed in IDH1/2 mutations, decreasing the levels of cis-aconitate, α–KG, malate and fumarate and increasing biosynthetic intermediates [352]. Changes in citric acid cycle intermediate levels in IDH1/2 mutants could affect the concentration of acetyl CoA and therefore contribute to an acetylation process that is critical for the activity of various tumorigenic proteins. 2-HG was found to alter PDH and pyruvate carboxylase activity which was also associated with levels of HIF-1α [353-355]. Cancer cells rely on aerobic glycolysis (the Warburg effect) during cell proliferation to accommodate the high demand for energy and to generate intermediates for biosynthesis of macromolecules such as nucleotides, proteins and lipids. A significant decrease in PDH activity in a 2-HG dependent manner was observed for an IDH1 mutant, contributing to the switch from oxidative phosphorylation to glycolysis. A higher level of HIF-1α in an IDH1 mutant resulted in elevated expression of PDK, causing a decrease in PDH activity.

Isocitrate Dehydrogenase and Cancer Treatment and Diagnosis

IDH1 and IDH2 may serve as good therapeutic targets owing to both being mutated in several human cancers. Several attempts have been made to identify inhibitors of IDH in an effort to develop a cancer therapeutic. AGI-5198, a compound which specifically inhibits TS603 glioma cells harboring the IDH1 Arg132His mutant in dose-dependent manner was identified. The compound also suppressed the colony formation of TS603 glioma cells on soft agar by 40–60%. Furthermore, AGI-5198 inhibited the growth of human glioma xenografts in mice by impairing tumor cell proliferation [356]. A small molecule, AGI-6780, that selectively inhibited an IDH2 mutant reduced 2-HG levels and promoted cell differentiation [357].

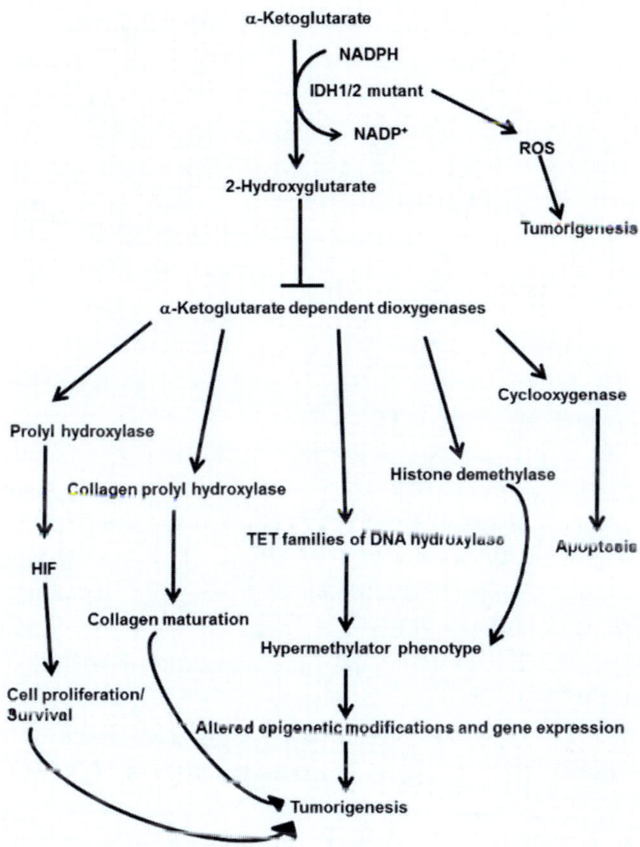

Figure 27. Potential mechanisms of tumorigenesis induced by IDH mutations. 2-hydroxyglutarate produced by IDH mutants can inhibit α-ketoglutarate-dependent dioxygenases, including prolyl hydroxylase, collagen prolyl hydroxylase, TET families of DNA hydroxylase, histone demethylase and cyclooxygenase. Suppression of TET families of DNA hydroxylase and histone demethylase causes hypermethylation, resulting in aberration in epigenetic modifications and gene expression. Inhibition of prolyl hydroxylase activates HIF, a transcription factor responsible for angiogenesis, apoptosis, metastasis and pathways involving in tumor growth. Reduction of collagen prolyl hydroxylase causes impairment of collagen protein maturation, contributing to embryonic lethality.

Following AGI-6780, novel oral inhibitors AG-120 and AG-221 which are specific for IDH1 and IDH2 mutations, respectively, were evaluated for their efficacies [357]. AG-221 markedly decreased 2-HG in marrow, plasma, and urine in a primary human AML xenograft model carrying the *IDH2* mutation. Additionally, treatment with AG-221 also demonstrated a dose-dependent survival benefit and dose-dependent decrease in cell proliferation and differentiation. The first in human phase I clinical trial of AG-221 was initiated in September 2014. The preliminary results of clinical trial indicated that the maximum tolerated dose was not reached and treatment was well tolerated. Side effects such as nausea, diarrhea, fatigue, and fever were observed. A decrease of up to 98% in plasma 2-HG levels was noted [358]. Further evaluation of AG-221 in solid tumors is ongoing. AG-120 is also been tested in a Phase I clinical trial in advanced hematologic malignancies and solid tumors with an IDH1 mutation, including gliomas, cholangiocarcinomas, and chondrosarcomas. Initial results were similar to those of AG-221 with adverse effects being differentiation syndrome, tongue edema and QT prolongation.

Another inhibitor of IDH1 mutants (Arg132His and Arg132Cys), ML309, was discovered using biochemical, cellular and biophysical approaches [359]. ML309 is a competitive inhibitor of α-KG in which it binds to the IDH1 Arg132His-NADPH complex. Moreover, in an attempt to identify a new selective chemical inhibitor of IDH1 Arg132His from a fragment-based chemical library, 2-(3-trifluoromethylphenyl)isothioazol-3(2H)-one was discovered, although the mechanism of action of this compound is under investigation [360]. A bioactive compound derived from *Withania somnifera*, Withaferin A, was also identified and found to exert its chemopreventive activity via IDH1 inactivation and mitochondrial dysfunction in early tumorigenesis [361].

The concept of synthetic lethality has been applied to therapy development [362]. This strategy is based on the non-oncogene addiction that cells carrying an oncogenic mutation exhibit significant dependency on a subset of non-oncogenes for survival. A functional RNA interference (RNAi) screen in AML cell lines expressing either IDH1 Arg132His or wild-type together with a green fluorescent protein marker was performed to identify non-oncogenes that are synthetic lethal to the IDH1 Arg132His mutant. It was revealed that two anti-apoptotic Bcl-2 family member genes, BCL2L2 or Bcl-w and Bcl-2, confer the strongest synthetic lethality. These two genes were found to be critical for cell survival in the IDH1 Arg132His mutant with elevated 2-HG levels. A further study focusing on Bcl-2 showed that knockdown of Bcl-2 by shRNA strongly decreased cell viability in the IDH1 mutant but not in the wild-type. The degree of Bcl-2 suppression was also found to be associated with the extent of apoptosis. Interestingly, IDH1/2 mutant primary human AML cells (IDH1 Arg132His, IDH 2 Arg140Gln an IDH2 Arg172Lys) were found to be more sensitive to inhibition by ABT-199, a highly specific inhibitor of Bcl-2 compared with normal cell lines.

The inhibitory effect of ABT-199 was further confirmed by an *in vitro* study in 39 AML patients and an *in vivo* experiment in which primary mutated IDH1/2 grafts leukemia cells were transplanted into mice, suggesting that ABT-199 confers a synthetic lethality. The mechanism by which ABT-199 contributes to Bcl-2 inhibition was investigated. It was also found that 2-HG, the product of IDH1/2 mutations, inhibited cytochrome c oxidase (COX), a result that was consistent to what was observed in AML cells in which IDH1/2 mutations caused a reduction in COX activity. Therefore, it is possible that combination of ABT-99 treatment and a COX inhibitor may provide a novel therapy for AML.

While small molecules can inhibit the production of 2-HG by IDH mutants, an immunotherapeutic approach targeting IDH1 mutations has been established as an alternative method for cancer treatment. Because IDH1 mutations occur early in tumorigenesis and are homogeneously presented within tumor cells, they are an attractive immunotherapeutic target in tumors. It was suggested that an immunotherapy method confers a great advantage in cancer treatment because this approach can abolish the target cells expressing the IDH1 mutation. $CD4^+$ T cells specific to IDH1 Arg132His were discovered and it was revealed that these $CD4^+$ T cells mediated antitumor immunity and growth suppression in gliomas [363].

2-HG was proposed as a potential biomarker of cancers harboring IDH1/2 mutations. Significant increased levels of 2-HG, a direct product of IDH1/2, was observed as a common feature of tumors with IDH1/2 mutations and its levels were also associated with the presence of IDH mutations in AML and glioma [311, 364]. 2-HG is released into serum, highlighting its potential as serum biomarker of cancers. Supporting this notion, the presence of IDH mutations in 18 AML samples was accurately identified based on levels of 2-HG [365]. Additionally, the presence of IDH2 Arg140Glu was discovered by monitoring the levels of 2-HG. However, the potential of using serum 2-HG as a predictive biomarker still requires careful evaluation. Magnetic resonance spectroscopy (MRS), a non-invasive approach, could be employed to detect the levels of 2-HG [366]. Alteration in PDH activity was observed in IDH1 mutant tumors and this activity could be detected by a technique such as MRS.

Therefore, it was also suggested that PDH activity could be served as therapeutic target for IDH1 mutant tumors [354]. MRS is useful for non-invasive diagnosis as well as for directing therapy before invasive treatments such as surgery.

Additionally, IDH mutations were proposed as a prognostic biomarker of prognosis in glioma, myeloid malignancies and other cancers. The impact of the presence of IDH1 mutation and MGMT methylation on long-term survival in low grade glioma patients was investigated [367]. It was revealed that 72.9% of patients carried an IDH1 mutation and this mutation was found to be closely associated with methylation in MGMT. The overall survival of initial surgery was positively correlated with the presence of IDH1 mutation and MGMT methylation. The meta-analysis (total of 20 studies included) of glioblastoma (GBM) patients revealed that GBM patients carrying an IDH1 mutation has decreased risk of mortality when compared with patients without an IDH1 mutation. The association of IDH1 mutation and reduced mortality risk is more profound in the European population than that in Asian citizens [368]. A mechanism by which an IDH1 mutation confers the non-aggressive features of glioma was proposed [369]. It was reported that the IDH1 Arg132His mutation inhibited the citric acid cycle and increased lipid metabolism. SREBPs were enhanced in U87 human GBM cells transfected with the IDH1 Arg132His mutant. Additionally, the presence of IDH1 mutation also decreased phosphorylation of retinoblastoma (Rb) protein. SREBPs are involved in the regulation of fatty acid and cholesterol biosynthesis as well as in the control of the CDK inhibitor p21 that arrests the cell cycle at G1. Therefore, it was proposed that the IDH1 mutation enhanced p21 expression via SREBPs and inhibited Rb phosphorylation, halting cell cycle progression. Additionally, it was hypothesized that IDH mutations confer a protective mechanism in glioma patients [370]. In contrast to other hypotheses, it was proposed that IDH mutations are not the origin of glioma but rather a subsequent mechanism to protect the cells, interfering with the metabolic pathway and causing cancer cells to be fragile and more susceptible to cell death.

In myeloid neoplasms, it was reported that IDH1 and IDH2 mutations increased the risk of disease severity, indicating that IDH mutations are involved in the clinical progression of the disease to a more advanced stage [316]. Moreover, IDH mutations were found to be correlated with overall survival, particularly in lower risk myelodysplastic syndromes. However, the impact of IDH1 and IDH2 mutations on treatment outcome was different. Therefore, the prevalence of IDH1 and IDH2 mutations should be considered separately in reference to their implications for diagnosis, prognosis and treatment. While IDH1 mutations are predominantly found in gliomas (95%), chondrosarcoma (95%) and cholangiocarcinoma (80%), IDH2 mutations have a higher frequency in AML. This could account for the variations in treatment observed in myeloid neoplasms.

In a study of 358 cytogenetically normal (CN)-AML patients, the IDH1 mutations Arg132His, Arg132Cys, Arg132Ser, and Val71Ile were found in 14% of patients and showed a prognostic impact on the subsets of younger molecular low-risk CN-AML patients [308]. Patients with IDH1 mutations had significantly shorter disease free survival and overall survival periods. IDH2 Arg140 mutations (Arg140Gln, Arg140Trp, Arg140Leu, and Arg140Lys) were found in CN-AML patients but conferred no prognostic significance. Interestingly, IDH2 Arg172 mutations are mutually exclusive with other prognostic mutations analyzed and signify a novel subset of older CN-AML patients characterized by resistance to induction chemotherapy. In contrast, the evaluation of gene mutations in IDH1 and IDH2 in 1,185 AML patients revealed none were statistically significant for overall survival [371]. Consistent with this previous report, overall survival was found to be independent on IDH mutations in a cohort study of 826 AML patients [358]. Though IDH1 mutations were not correlated with prognosis in AML, the expression of IDH1 is associated with a poor prognosis of CN-AML in a study of 320 patients using quantitative real-time reverse-transcription (RT)-PCR [372].

In a HCC cohort of 419 patients, *IDH* gene polymorphisms (rs12478635 in *IDH1* and rs11632348 in *IDH2* gene) exhibited strong association with the death risk in HCC patients treated with transcatheter arterial chemoembolization in a dominant model [373]. Furthermore, proteomic analysis suggested that increased expression of IDH1 could be used as a plasma biomarker for diagnosis and a histochemical biomarker for prognosis prediction of NSCLC [374].

In addition to cancers, IDH1 was found to play an important role in other diseases. IDH1 was reported to be a negative regulator of glucose signaling for insulin secretion in pancreatic β-cell [375]. Knockdown of IDH1 by RNAi in INS 832/13 cells resulted in enhancement of glucose-induced insulin secretion (GIIS) but did not disrupt glucose oxidation. However, inhibition of IDH1 was found to reduce fatty acid oxidation and increase lipid biosynthesis from glucose. Metabolomic analysis of INS 832/13 cells indicated increased isocitrate and $NADP^+$ levels after IDH1 suppression. Levels of other metabolites related to GIIS were also increased, including acetyl CoA, glutamate, cAMP and ATP. Therefore, it was suspected that IDH1 could contribute to the health condition of diabetics.

In addition to IDH1 and IDH2, IDH3 has been found to play an important role in many diseases. In bipolar disorder (BD), metabolomic analysis of the cerebrospinal fluid (CSF) from 54 Swedish male BD patients revealed that the levels of isocitrate in CFS from BD patients were significantly higher than those of healthy controls [376]. The mechanism underlying this observation was that expression of two subtypes of *IDH3* gene (*IDH3A* and *IDH3B*) in dorsolateral prefrontal cortex from BD patients was significantly lower than that of controls. Accordingly, protein expression of IDH3A in the cerebellum of BD patients was also lower than that of the controls. However, no genetic alterations in the *IDH3* gene were observed, suggesting that genetic variation was not associated with BD in the Swedish population. The data suggested that low expression of IDH3 contributing to mitochondria dysfunction may be involved in the pathogenesis of BD. However, the sample size in this study is small. Therefore, further studies with larger sample size are required to confirm the pathogenic role of IDH3 in BD.

Furthermore, IDH3 was reported to be involved in cancer-associated fibroblasts [377]. miR-424 downregulated IDH3α in cancer-associated fibroblasts. Decreased IDH3α indicates the switch from oxidative phosphorylation to aerobic glycolysis in cancer-associated fibroblasts. The reduction in IDH3α led to diminished levels of 2-KG and resulted in inhibition of PDH2 and stabilization of HIF-1α. Reduced HIF-1α was also observed in cancer-associated fibroblasts from tumor samples, indicating the role of IDH3α as a metabolic switch in cancer-associated fibroblasts.

α-Ketoglutarate Dehydrogenase Complex

α-Ketoglutarate dehydrogenase complex (KGDHc; EC 1.2.4.2) is a key enzyme in the TCA cycle that catalyzes the conversion of α-KG into succinyl CoA. Similar to the PDHc, this enzyme is a multienzyme composed of three subunits: α-ketoglutarate dehydrogenase (E1), dihydrolipoamide succinyltransferase (E2), and dihydrolipoamide dehydrogenase (E3). KGDHc reaction is started by the oxidative decarboxylation of α-KG by E1, resulting in the formation of a succinyl group that is attached to thiamine pyrophosphate. The succinyl group is then transferred to lipoic acid on the E2 subunit. Next, E2 catalyzes the transfer of the succinyl group from the lipoate moiety to CoA, resulting in the formation of succinyl CoA and dihydrolipoate. E3 is a flavoprotein shared with PDHc that catalyzes the transfer of reducing equivalents from dihydrolipoate to produce NADH, and recycling dihydrolipoate to lipoate that participates in the next reaction cycle, Figure 28 [378].

KGDHc is regulated by several substrates, cofactors, and products. The major factors that regulate the activity of KGDHc are the CoA/succinyl CoA, NAD^+/NADH, and ADP/ATP ratios. Moreover, some divalent cations such as Mg^{2+} and Ca^{2+} modulate KGDHc activity by reducing the K_m for α-KG [378, 379]. KGDHc is not only important in the TCA cycle, but also has a key role in other mitochondrial metabolic pathways such as glutamine catabolism. Glutamine is the most abundant amino acid in the circulation and intracellular fluid. It is an important precursor for the synthesis of many biological macromolecules. Glutamine can be metabolized into α-KG, which is further catalyzed by KGDHc in the TCA cycle. Therefore, defects in KGDHc can affect glutamine metabolism [380].

α-Ketoglutarate Dehydrogenase Complex and Oxidative Stress

Oxidative stress, a condition where excessive ROS are produced and secreted, has been identified as a contributor to various human pathologies, especially morbidity in the nervous system. Previous studies revealed that oxidative stress mediates mitochondrial dysfunction, as indicated by a decline in ATP and NADH levels, and changes in the mitochondrial membrane potential. Because the structure of KGDHc contains a large number of sulfhydryl groups and tyrosines that are targets for free radicals, it is sensitive to ROS [381]. Based on this, it was hypothesized that one of the underlying mechanisms of oxidative stress induced cell damage is the regulation of KGDHc activity.

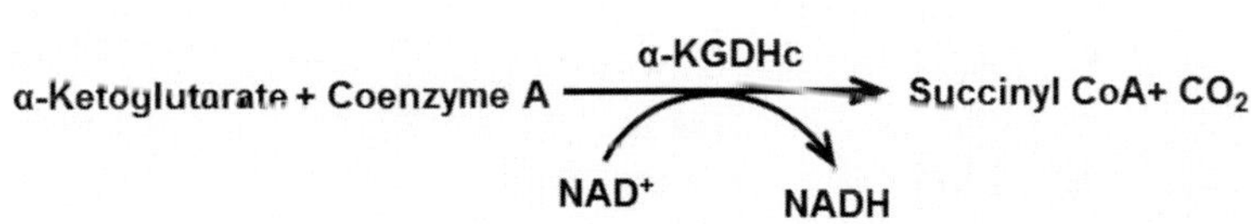

Figure 28. Reaction catalyzed by α-KGDHc.

The effect of oxidative stress causes KGDHc inactivation as demonstrated in H_2O_2-induced oxidative stress. The nerve terminals of guinea pigs were treated with H_2O_2, followed by the investigation of TCA cycle enzymes and changes in mitochondrial metabolism. It was found that three enzymes, aconitase, KGDHc, and SDH, were inhibited during the acute expose of nerve terminals to H_2O_2. In addition, decreased NAD(P)H was observed proportional to the oxidant concentration without a change in total amount of the pyridine nucleotide pool. To establish which enzyme limits NAD(P)H production during H_2O_2-induced oxidative stress, the relationships between decreased NAD(P)H levels and the activity of each enzyme were analyzed. The inhibition of KGDHc correlated with decreased NAD(P)H levels, suggesting that limited NAD(P)H production in the TCA cycle during exposure to oxidative stress is caused by the inhibition of KGDHc [382]. This result was also demonstrated in a study of heart mitochondria induced by H_2O_2 [383].

Several toxic products are produced during oxidative stress. 4-hydroxy-2-nonenal (HNE), an aldehydic product of lipid peroxidation, exhibits an effect on rat heart mitochondrial function by decreasing the production of NADH. No change in NADH-related enzymes was observed except for a reduction of KGDHc, indicating that reduced NADH may be caused by HNE-mediated KGDHc function. To determine the mechanism of inhibition, the effects of HNE on purified KGDHc were examined and showed that HNE reduced the sulfur atoms of lipoic acid covalently bound to the E2 subunits, interrupting the binding between substrate and enzyme. In addition, western blot analysis using a native lipoic acid antibody revealed the loss of antibody recognition in HNE-treated KGDHc, providing the support for the modification of KGDHc by HNE [384].

Glutathione has recently been recognized as a regulatory molecule that can form a reversible disulfide bond with a sulfhydryl group on other proteins, resulting in the regulation of protein structure and function. Previously, it was reported that this process is involved in cellular responses under various stress conditions. Applegate et al. performed experiments in rat heart mitochondria to determine whether glutathionylation blocks the inactivation of KGDHc by the sulfhydryl reactive compound N-ethylmaleimide (NEM). Reactivation of H_2O_2-treated enzymes was observed for medium supplemented with glutaredoxin and GSH. In addition, deglutathionylation enhances the susceptibility of the enzyme to NEM. Thus, glutathionylation prevents enzyme modification by NEM. Further experiments revealed that the E2 subunit is glutathionylated in response to H_2O_2, suggesting that lipoic acid represents the likely site of glutathionylation. In summary, this study demonstrated that the glutathionylation of KGDHE2 is an antioxidant response that protects the enzyme from oxidative damage [385].

Although evidence demonstrating KGDHc might be a crucial target of ROS has already been reported, KGDHc can also act as a generator of oxidative stress. Initially, it was thought that mitochondria were a source of ROS production. Many studies have supported the idea that ROS are mainly generated by processing of the mitochondrial electron transport chain. Furthermore, it was reported that flavoproteins are involved in oxygen activation, resulting in the production of superoxide and hydrogen peroxide. The E3 component of KGDHc is a flavoenzyme that transfers electrons from reduced lipoamide, which contains an E2 subunit, to NAD^+, the final electron acceptor in a KGDHc catalyzed reaction. Tretter and Adam-Vizi demonstrated that KGDHc could generate H_2O_2 [386].

Nerve terminals isolated from the brain cortex of guinea pigs were incubated in medium containing no substrate, glucose, or α-KG. H_2O_2 in the medium was detected using the Amplex Red fluorescent dye. A remarkable increase in H_2O_2 generation was observed in the presence of α-KG. However, no change in H_2O_2 levels was found in the presence of glucose. These results indicated that H_2O_2 production might be increased by KGDHc. Moreover, enhanced H_2O_2 can be inhibited by NAD^+ and stimulated by NADH. Based on the opposite effect of NAD^+ and NADH on KGDHc mediated H_2O_2 generation, the dependence of H_2O_2 formation on the NADH/NAD^+ ratio was investigated. It was found that an increase in the NADH/NAD^+ ratio significantly stimulated H_2O_2 generation by KGDHc. Taken together, it was concluded that KGDHc is not only a target of ROS but might also contribute to the generation of oxidative stress when NADH oxidation is impaired [386].

An additional study by Starkov et al. demonstrated KGDHc could generate ROS. Heterozygous knock-out mice deficient in dihydrolipoyl dehydrogenase (DLD), known as the KGDHE3 subunit, were established to verify the results from an *in vitro* study. Although DLD is shared between PDHc and KGDHc, no significant changes in the respiratory chain and maximum respiratory rate of other TCA components were observed except for α-KG. However, brain mitochondria isolated from DLD-knockdown mice produced significantly less H_2O_2 than mitochondria isolated from wild-type mice [387].

Other studies using cultured hippocampal neurons from rats revealed that KGDHc also contributes to the production of ROS in glutamate-stimulated neurons. Enhanced neuronal ROS generation was inhibited by the addition of a specific KGDHc inhibitor [388]. Overall, these studies strongly indicated that the flavoenzyme KGDHE3 subunit functions as an oxidative stress generator in mitochondria.

α-Ketoglutarate Dehydrogenase Complex and Neurodegenerative Diseases

Oxidative stress is a common feature of many age-related neurodegenerative diseases. Previously, a putative mechanism whereby oxidative stress leads to neuron damage via the inactivation of KGDHc activity was described. Next, how KGDHc dysfunction can cause brain pathogenesis will be discussed.

One way that a reduction of KGDHc can impair the regular function of the brain is altered energy metabolism. Mice cerebellar granule neurons were used to investigate the relationship between reduced KGDHc and glucose utilization. Cells were treated with a specific inhibitor of KGDHc and changes in [1-^{13}C] glucose and [U-^{13}C] glutamate metabolism were determined. A decreased formation of intermediates derived from [1-^{13}C] glucose was observed. Moreover, the decreased formation of intermediates derived from [U-^{13}C] glutamate was also observed along with a decline of the concentrations of several amino acids. These results indicated an impaired flow of glucose and glutamate carbon through the TCA cycle. According to the significance of glucose as an energy source and the role of glutamate in neurotransmitter synthesis, this alteration might lead to the abnormal function of the nervous system [389].

A related study was performed using mice with the disrupted expression of dihydrolipoyl succinyltransferase (DLST, also known as the E2 subunit). Brain tissues were extracted from the cortex and cerebellum and then injected with [U-^{13}C] glucose. An increase in the concentration of glucose in the cortex was observed, suggesting a decrease in glucose utilization in DLST knockdown mice [390]. Comparing the metabolism of [U-^{13}C] glucose in a human embryonic kidney cell line with a cell line where the KGDHc activity was about 70% of the control revealed increased labeled lactate, indicating the flux of metabolism into glycolysis [391]. This change may contribute to lactic acidosis, a clinical manifestation usually found in KGDHc deficient patients [392, 393].

Another effect of KGDHc deficiency on brain pathogenesis is selective cell death. Cell lines treated with a KGDHc inhibitor were used to test the effects of reduced KGDHc on cell death signaling cascades. Enhanced translocation of mitochondrial cytochrome c into the cytoplasm was observed along with activated caspase-3 and a change in mitochondrial permeability. Therefore, this suggests that the inhibition of KGDHc mediates the permeability of the mitochondrial membrane. This alteration facilitates the release of cytochrome c from mitochondria, and activation of the caspase signaling pathway that participates in cell apoptosis induces cell death [394].

A further study was performed in KGDHc E2 and E3 subunit deficient mice. Immunohistochemistry for immature neuron markers, a neural cell adhesion molecule, as well as markers for proliferation showed reduced doublecortin-positive neuroblasts in the brain tissue of E2 and E3 subunit knockdown mice [395]. This result indicated a selective reduction in the number of progenitor cells in the brain, which may be influenced by cell elimination or proliferation inhibition caused by KGDHc deficiency. Interestingly, KGDHc was recently reported to have a role in protein modification. Post-translational modification is a process that affects protein function. Therefore, abnormal KGDHc might disturb the regular function of other proteins. Succinylation is a modification where a succinyl group is added to a lysine residue on a protein. E2 is a succinyltransferase and its downstream product is succinyl CoA. Weinert et al. performed experiments to test whether KGDHc succinylated proteins without the need for other enzymes. KGDHE1 deficient yeast had a 6-fold reduction in global succinylation [396]. These results indicated that KGDHc participates in the succinylation of protein.

A later study demonstrated that neuron cell lines treated with a KGDHc specific inhibitor showed significantly reduced succinylation in cytosolic and mitochondrial proteins. Remarkably, acetylation was also sensitive to the inhibition of KGDHc. To test whether KGDHc can donate succinyl groups to other proteins, KGDHc was incubated in the presence and absence of succinyl CoA and α-KG. It was found that strongest succinylated proteins were observed in the presence of KGDHc and α-KG, while succinyl CoA provided only a slight increase in succinylation. Taken together, these findings provide evidence that KGDHc has a role in cellular functions by regulating protein modification [397].

In summary, there are three possible underlying mechanisms involved in KGDHc deficient neuropathogenesis: (1) metabolic change, (2) selective cell death, and (3) alterations in post-translational modification. The major neurodegenerative diseases associated with KGDHc function are AD and PD. Next, the association between the pathogenesis of AD and PD and a loss of KGDHc will be discussed.

AD is a chronic neurodegenerative disease where the most common symptom is difficulty in remembering recent events. A large number of studies reported a reduction of 45–75% in KGDHc activity in the brains of AD patients [398, 399]. To determine whether this is associated with an abnormality in one of the three constituent enzyme subunits, Mastrogiacoma et al. measured the protein levels of El, E2, and E3 subunits in the postmortem brain of 29 patients with AD and 29 control subjects. In the AD group, protein levels of all three subunits were significantly reduced by 23–41% in the temporal cortex, whereas in the parietal cortex (El: 28%; E3: 32%) and hippocampus (E3: 33%) significant changes were limited to El and E3. Remarkably, reduced KGDHc did not correlate with protein levels, suggesting that decreased enzyme activity cannot be primarily explained by the loss of KGDHc protein [398].

KGDHc activity has been associated with the genotype of *DLST*, the gene that encodes the E2 subunit. Polymorphisms of *DLST* associated with AD are A19117G in intron 13 and T19183C in exon 14. Homozygous G and C alleles of *DLST* enhanced the association of APOE4 with AD in 429 white Jewish patients [400]. Moreover, homozygous A and C alleles were associated with AD in 251 Japanese patients [401]. However, no association between *DLST* gene polymorphism and AD was observed in a study by Matsushita et al. [402].

A study of the association between genotype and KGDHc phenotype was also performed for other subunit genes including *DLD*, the gene that encodes the E3 subunit. No association was observed between genotype in *DLD* and AD, as each variant occurred less frequently. The combination of gender and *DLD* genotype had a significant association with AD in both Caucasian and Ashkenazi Jewish patients, but this was only apparent in the male population [403]. Although mutations of the gene encoding KGDHc does not directly contribute to brain pathogenesis, genetic variations in the *KGDH* gene may increase the risk for AD development.

PD is a movement disorder characterized by tremors, rigidity, slowness of movement, and postural imbalance. The primary pathologic abnormalities are degeneration of the substantia nigra and dopaminergic neurons of the striatum, which leads to a decrease in dopamine levels [404, 405]. Evidence suggests that KGDHc declines in PD. Different from AD, a loss of KGDHc in PD usually occurs by the inhibitory effects of neurotoxins. 1-methyl-4-phenyl-1, 2, 3, 6-tetrahydropyridine (MPTP) or 1-methyl-4-phenylpyridinium (MPP^{+}) are present at high levels in the brain of PD and were reported to inhibit the activity of KGDHc. However, these toxins inhibit KGDHc through a radical-mediated mechanism rather than through a direct interaction with KGDHc [406]. An immunohistochemical study demonstrated reduced KGDHc in the substantia nigra of PD patients. Furthermore, KGDHc immunostaining was decreased in many melanized neurons, and this reduction correlated with the severity of degeneration [407].

An association between the *KGDHc* gene and PD was studied by Kobayashi et al. A biallelic intragenic polymorphism was identified in the *E2* gene of *KGDHc*. It was a single nucleotide substitution between G (in allele 1) and A (in allele 2) at a position that does not change the amino acid code. Frequencies of genotypes that carry allele 2 were significantly higher in the PD group than in the control group. These results indicated that a genetic variant of the *E2* gene itself or in close proximity to the gene constitutes a genetic risk factor for PD [408].

α-Ketoglutarate Dehydrogenase Complex and Cancer

KGDHc is often implied to be inactive in cancer. Quantitative PCR analysis showed an 8-fold downregulation of *KGDHc* in 50% (15/30) of colorectal cancer samples. Bisulfite sequencing revealed hypermethylation of the *KGDHc* promoter region, suggesting the reduction in KGDHc is mediated by an epigenetic alteration. Based on this finding, the *KGDHc* promoter methylation pattern might provide a marker for the differential diagnosis of $CIMP^+$ tumors, which display dense hypermethylation of the promoter region of many genes [409].

Different cancer types show a reduction in the expression of KGDHc indicating that the activation of KGDHc may modulate tumor growth. However, the molecular mechanism by which KGDHc exerts its growth modulating function has not been elucidated. Sen et al. demonstrated a role for overexpressed KGDHc in cervical cancer growth. The restoration of KGDHc expression in cervical cancer cells lacking endogenous KGDHc expression suppressed cell proliferation, invasion, and soft agar colony formation. In contrast, the knockdown of KGDHc expression in cervical cancer cells expressing endogenous KGDHc had the opposite effect. Moreover, the overexpression of KGDHc increased the production of ROS, leading to apoptosis through caspase 3 activation. This effect mediated the down-regulation of the Akt signaling cascade and decreased NF-kB phosphorylation, a signal transduction pathway that promotes cell survival and growth. Taken together, these data suggest that the downregulation of KGDHc can contribute to tumorigenesis; therefore, the induction of KGDHc may be a potential therapeutic strategy for cancer treatment [410].

Not all cancers prefer the conditions in which KGDHc is downregulated. Bunik et al. investigated the role of cancer cell homeostasis in various cancer cell lines by adding succinyl phosphonate, an inhibitor of KGDHc. Different responses were observed in different cancer cell lines. Some cancers were sensitive to succinyl phosphonate treatment, as indicated by reduced cell viability, decreased NAD(P)H and ATP levels, and increased glutamate and ROS production [411]. These results indicated that some cancers maintain oxidative phosphorylation as an energy source. KGDHc therefore participates in cancer survival and KGDHc inhibition might have diagnostic value for anticancer strategies. The role of KGDHc in disease development is illustrated in Figure 29.

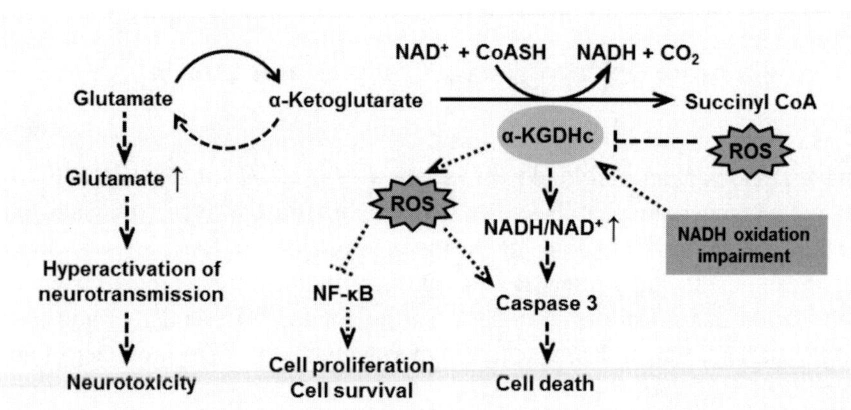

Figure 29. The role of α-KGDHc in disease development. KGDH is sensitive to oxidative stress. Inactivation of KGDH results in change of redox state, activating the caspase-induced apoptotic pathway. KGDH is also associated with glutamate metabolism, loss of enzyme function leads to the accumulation of glutamate, resulting in the neurotoxicity. In addition, KGDH is able to generate ROS in the absence of NAD^+, an electron acceptor. This oxidative stress inhibits pro-apoptotic pathway and enhances cell apoptosis.

Succinate Dehydrogenase

Succinate dehydrogenase (SDH; EC 1.3.5.1), also known as a complex II, plays an important role in both the TCA cycle and electron transport chain. It comprises four subunits, SDHA and SDHB, which face the mitochondrial matrix, and SDHC and SDHD, which are embedded in the mitochondrial inner membrane. SDHA catalyzes the conversion of succinate into fumarate. In this reaction, electrons are also oxidized by FAD to form $FADH_2$ (Figure 30). Then, electrons from $FADH_2$ are transferred to the iron cluster in the SDHB subunit. The electrons fed to SDHB are then transferred to the ubiquinone pool in the SDHC/SDHD dimer, which then shuttles electrons to the downstream respiratory chain complexes (Figure 31) [412]. In addition to four subunits, there are accessory factors, SDHAF1 and SDHAF2, which participate in SDH function. Although the functional roles of these two factors remain poorly understood, several studies suggested that SDHAF1 and SDHAF2 might play a role in the insertion of iron clusters and FAD in the SDH complex, respectively [413].

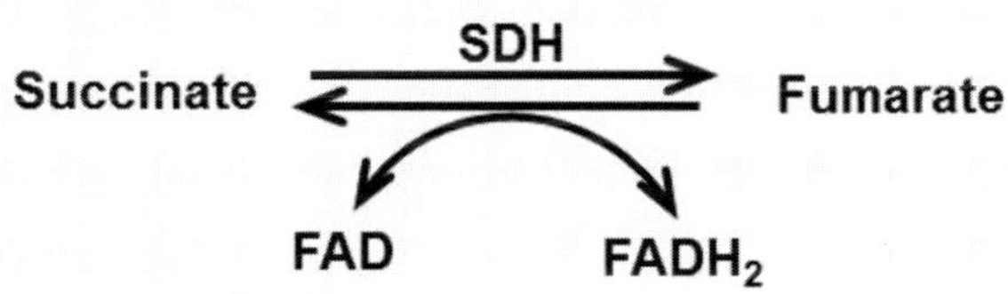

Figure 30. Reaction catalyzed by SDH.

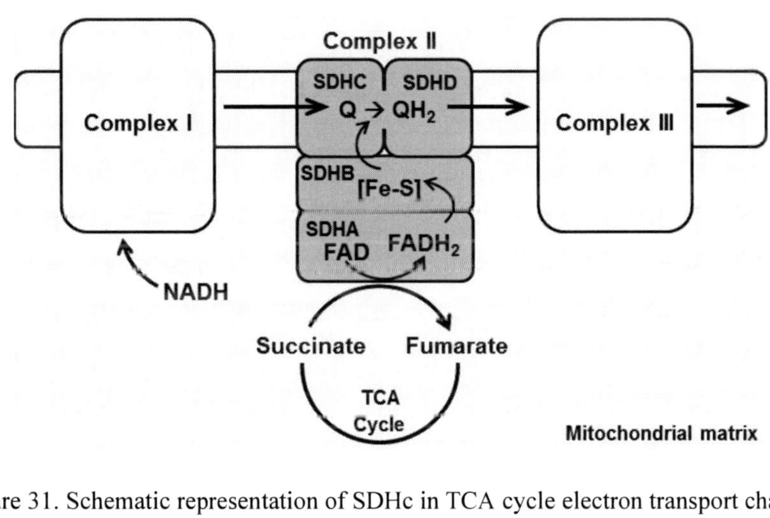

Figure 31. Schematic representation of SDHc in TCA cycle electron transport chain.

Defects in SDH and its accessory factors lead to improper functioning of the TCA cycle and oxidative phosphorylation, and for that reason, many mutations in genes encoding this enzyme have been demonstrated in various pathologies. Accordingly, the associations between SDH and diseases, including several cancers, neurodegenerative diseases, as well as cardiovascular diseases, are described in this review.

Succinate Dehydrogenase and Cancer

Paragangliomas (PGLs) are neuroendocrine tumors that originated from the autonomic nervous system paraganglial. PGLs mainly arise in the head and neck, and are usually slowing growing benign tumors that rarely metastasize, whereas pheochromocytomas (PCCs), described as PGLs that arise in the adrenal gland, are often malignant, and are involved in hypertension crisis via the hypersecretion of catecholamines and other hormones [413, 414]. Most PGLs are sporadic, although some cases associated with germline mutations in several susceptibility genes, including *RET*, *VHL*, and *NF1* [414], are hereditary. The PGLs are also caused by germline mutations in genes encoding four subunits and the assembly factor of the SDH complex. PGLs can be classified into four clinical entities, PGL1, PGL2, PGL3, and PGL4, which correspond to specific germline mutations in the *SDH* gene (Table 5).

Table 5. Summary of SDH –related PGL/PCCs

Clinical entitle	Clinical feature	The common SDH mutation				Reference
		Gene	Location	Nucleotide change	Protein change	
PGL1	Multifocal PGLs	*SDHD*	Exon 3	c.274G > T	p.Asp92Tyr	[415-417]
PGL2	Multifocal PGLs at young	*SDHAF2*	Exon 2	c.232G > A	p.Gly78Arg	[418-420]
PGL3	PCCs	*SDHC*	Intron 5	c.IVS5 + 1G > T, A	Frameshift	[421, 422]
PGL4	Metastatic PCCs	*SDHB*	Exon 2	c.136C > G, T	p.Arg46Gly, X	[415, 423, 424]

PGL1 is caused by the mutations in the *SDHD* gene. Familial PGLs caused by germline mutations in the *SDHD* gene generally develop multifocal head and neck paragangliomas (HNPGLs), and less commonly adrenal PCCs and metastases. Although malignant tumors rarely occur in PGL1, *SDHD* nonsense mutations were reported as a significant risk factor for PCCs [415-417].

In the study by Ricketts et al., from a total of 358 PGLs patients, 63 patients were identified with 10 different *SDHD* mutations. To assess the genotype-proteotype-phenotype correlation, the mutations were classified into two groups, nonsense and missense mutations. It was revealed that patients with loss of expression or truncated SDHD have a higher risk to develop PCCs than those with *SDHD* missense mutations. The mutation of exon 3, c.242C4T; p.Pro81Leu, was most commonly found in *SDHD* carriers, and was not previously described. A structural prediction analysis of SDHD revealed that a p.Pro81Leu missense substitution leads to changes in protein folding and destroys the ubiquinone binding site, which in turn might result in less functional SDH [424]. Based on these data, SDHD-defected PGL1 mainly forms HNPGLs and is currently associated with the development of PCCs.

PGL2 is caused by a mutation in the *SDHAF2* gene. The association between the *SDHAF2* mutation and PGLs is very new and studies are ongoing. In a study of *SDHAF2*-mutated yeast, decreased SDH activity and expression, and flavo-SDHA were observed [418]. These data suggested that SDHAF2 is essential for the incorporation of FAD into the SDHA subunit, and thus mutation of *SDHAF2* results in the loss of SDH stability and function. The germline mutation in the *SDHAF2* gene, exon 2 c.232G > A; p.Gly78Arg, was previously found in PGL patients. First, it was reported in the Dutch kindred of familial HNPGLs. As observed in yeast, SDHA flavination was decreased by ~95% in tumors from PGL2 patients [418]. A second report investigated a Spanish family which developed HNPGLs at a young age [419]. Recently, a new branch of Dutch PGL2 was also reported in an HNPGLs family that showed the onset at young age and high multifocality [420]. Taken together, these results suggest that hereditary PGLs caused by a germline mutation in the *SDHAF2* gene are rare and mainly manifest in multifocal HNPGLs in young patients.

PGL3 is caused by a mutation in the *SDHC* gene. Because of the functional collaboration between SDHD and SDHC, familial PGLs caused by a *SDHC* mutation are similar to those caused by a *SDHD* mutation. However, cases of multiple and hypersecreted-PGL with a *SDHC* mutation seem to be more frequent than those with a *SDHD* mutation. Patients with malignant or catecholamine-producing carotid PGLs contained a germline mutation in the *SDHC* gene, a G>T transversion at position +1 of intron 5, which in turn leads to the deletion of exon 5 and a reading-frameshift of exon 6 [421]. A mutation on *SDHC* intron 5 was also found in cardiac PGLs, but the G at position +1 was traversed by A, c.IVS 5+1, G>A [422]. Later, a patient presented with abdominal PGL metastatic to the skeleton and multiple lymph nodes was determined to have a large allelic deletion of exons 1–6 in *SDHC* [425]. Moreover, a retrospective study of eight PGL3 patients revealed three of them had mediastinal PGLs that have elevated normetanephrine levels. Another four cases were multiple PGLs, those of which were thoracic PGLs had increased levels of normetanephrine, norepinephrine, and dopamine [426]. These data indicated that patients with PGL3 usually present benign HNPGLs or thoracic PGLs, and may also develop catecholamine hypersecreted-malignancies.

PGL4 is caused by mutation in the *SDHB* gene, which is the gene most commonly mutated of the *SDH*-related genes. In contrast with other clinical entities, patients with *SDHB*-related PGLs very frequently develop metastatic diseases and malignant PGLs. *SDHB* mutation carriers mainly have adrenal PCCs with hypersecretion of catecholamines. One germline mutation in *SDHB*, a C>T transversion at nucleotide 402 in exon 3, was reported in British kindred that developed extra-adrenal PCCs at young age, and in Swedish kindred with both extra-adrenal PCCs and cervical PGLs. This mutation leads to a change of arginine at position 91 to a terminator codon (Arg91stop), which in turn results in truncated SDHB. Another mutation, a C>G transversion at nucleotide 724 in exon 6, causing a missense mutation of proline to arginine (Pro198Arg), was also identified in familial extra-adrenal PCCs. *SDHB* mutations not only cause hereditary PCCs, but also are associated with sporadic disease, as shown by the deletion of C at nucleotide 725 in exon 6 (725delC) identified in sporadic adrenal PCCs [427]. *SDHB*-associated catecholamine hypersecretion was reported in a PGL4 carrier with carotid body PGLs. In this patient, laboratory results demonstrated elevations of metanephrines and norepinephrine. Moreover, metastatic lymph nodes and multiple PGLs were observed. After *SDH* gene analysis, this patient was found to have a mutation in *SDHB*, exon 7 c.689G > A, p.Arg230His [428]. Consequently, possession of a germline *SDHB* mutation should be considered a risk factor for the development of metastatic and severely malignant PGLs.

Germline mutations of *SDHA*, resulting in a loss of SDH function, have also been reported in PGLs. A patient affected with catecholamine-secreting abdominal PGLs was shown to have the *SDHA* mutation c.1765C.T. This mutation causes a change from arginine 589 to tryptophan, which results in

SDHA protein instability. Immunohistochemistry of a tumor from patient revealed pseudohypoxia and the absence of SDH, similar to what has been observed for patients with other *SDH* mutations [429]. Another patient with bladder PGLs was shown to have single base pair substitutions in exon 5 (c.484A > T; p.Arg162stop) and exon 10 (c.1340A > G; p.His447Arg) [430]. Furthermore, a novel *SDHA* mutation (c.1873C > T; p.His625Tyr) was found in the carotid bodies from a PGL patient as well as in his son who presented with pituitary adenoma [431]. Even *SDHA* mutations in PGLs are rare; defects of SDHA are usually associated with a neurodegenerative disorder known as Leigh syndrome which will be described below.

Gastrointestinal stromal tumors (GISTs) are mesenchymal tumors of the gastrointestinal tract that usually arise in the stomach and small intestinal and rarely occur elsewhere within the GI tract. Most GISTs are driven from the mutation of the oncogenic genes, *KIT* and platelet-derived growth factor receptor-α *(PDGFRA)*. Approximately 10–15% of GISTs do not have *KIT* or *PDGFRA* mutations and are referred to as wild-type GISTs. It has been reported that abnormalities of several genes, such as *NF1*, *BARF*, the *RAS* family, and *SDHx*, may contribute to the development of wild-type GISTs. Among 756 GISTs, 66 SDH-deficient GISTs (7.6%) were identified by loss of SDHB from immunohistochemical analysis, which indicates the SDH complex was ineffective in these cases. SDH-deficient GISTs vary in terms of tumor size and mitotic activity, but are associated with pediatric GISTs [432]. Pediatric GISTs often develop multifocal tumors and metastatic disease, two characteristics which are observed in adult GISTs. SDH-deficient pediatric GISTs were also demonstrated in a study of sporadic wild-type GISTs. It was revealed that GIST patients with *SDHx* mutations were all young adults, 16–22 years at diagnosis, and all had multifocal GISTs. To assess the relationship between pediatric GISTs and SDH, immunohistochemistry and western blotting studies of SDHB expression were performed in 18 and 12 cases that were identified as pediatric (age of diagnosis ≤18 years) and adult wild-type GISTs, respectively. SDHB protein expression was absent in all of the 18 (100%) pediatric cases, and absent in 8 of 12 (67%) and low in 4 of 12 adult cases [433]. These data suggested that loss of SDH function is involved in the development of wild-type GISTs, especially in pediatric patients. The associations between wild-type GISTs and SDH deficiency have been described in terms of Carney-Stratakis Syndrome (CSS) and Carney Triad (CT). CSS is characterized by the development of wild-type GISTs because of the mutations of *SDH* subunit genes, while CT describe wild-type GISTs caused by loss of SDH expression without mutations in any *SDH* genes [434].

CSS is inherited and can occur in both males and females. GISTs caused by mutations of *SDH* genes have been reported in several studies. In the study by Janeway et al. 36 patients with wild-type GISTs were subjected to *SDHB, -C,* and *-D* gene sequencing. Four patients had mutations in *SDHB* and *SHDC*, of which three of these mutations had been previously reported in PGLs. The novel *SDHB* mutation, exon 3 c.274T > A; p.Ser92Thr, was predicted to inactivate SDHB function, based on *in silico* analysis. To confirm the impact of this mutation, SDHB immunohistochemistry was performed and absent SDHB protein expression was observed [433].

Novel mutations in *SDH* genes were also reported by Pantaleo et al. Genome sequencing of all *SDH* subunits was performed in 34 wild-type GISTs. Nine patients were found to have mutations in *SDH*x (6 in *SDHA*, 2 in *SDHB*, and 1 in *SDHC*). From this study, the phenotypes of the non-synonymous mutations that have not been previously described were predicted by bioinformatics tools. The result revealed five novel mutations (SDHA: Arg171Cys, Gly419Arg, Glu564Lys, and Arg589Gln; SDHC: Glu144Gln) that have high probability of causing protein damage and are likely related to disease progression [435]. Other mutations were reported in the *SDHA* gene. Among 33 SDH-deficient GISTs, nine patients were identified with seven different *SDHA* mutations. The mutation of exon 2, c.91C4T; p. Arg31stop, was most commonly found in *SDHA* carriers, and also has been most frequently reported in *SDHA*-mutated PGLs [436]. Interestingly, this mutation may be a biomarker for screening *SDH*-related tumors.

In contrast with CSS, CT is not inherited and usually occurs in woman. Remarkably, these GISTs are frequently multifocal, and metastasize into lymph nodes as well as distant organs [437]. Although the etiology of SDH deficiency in CT remains unclear, several studies demonstrated that it may result from epigenetic alterations in *SDHx* genes. In the study by Haller et al. patients with CT, CSS, PGL1, KIT-GISTs were subjected to massive parallel bisulfite sequencing to evaluate CpG methylation in all *SDHx* subunits. DNA hypermethylation of the *SDHC* gene in tumors was observed along with reduced *SDHC* mRNA in CT patients, but was not present in other patients. Moreover, this pattern was correlated with the loss of SDHC and SDHB protein levels as well as SDH complex activity [438]. In another study, a case report of a wild-type GIST patient revealed the absence of immunohistochemical SDHB. Gene expression analysis of four SDH subunits revealed a remarkable reduction of *SDHC* mRNA along with hypermethylation of the CpG island on the *SDHC* promoter [439].

Taken together, it can be concluded that epigenetic modification of the *SDHx* gene leads to the downregulation of SDHx. Loss of SDHx can cause the conformational and functional impairment of the SDH complex, which in turn may be an underlying mechanism of SDH-deficient CT.

In addition to PGLs, PCCs and GISTs, rare cases of various tumors have been associated with mutations of *SDH* genes. Ricketts et al. revealed the germline mutations of *SDHB* in renal carcinoma patients, including c.32G > A; p.Arg11His in exon 1, c.136C > T; p.Arg46stop and c.137G > A; p.Arg46Gln in exon 2 [440]. Additionally, a case report of a man with renal carcinoma also revealed a mutation in *SDHB* exon 2, c.88delC; p.Gln30ArgfsX47. SDHB immunohistochemistry results demonstrated the complete absence of staining in tumor cells, but positive staining in normal cells [441]. It was reported that *SDH* mutations in renal carcinomas are usually found in the *SDHB*, *-C*, and *–D* genes, and approximately 30% of affected patient may develop multiple tumors [442]. Moreover, SDHB expression was correlated with survival of renal carcinoma [443].

SDH-related PGLs usually coexist with pituitary tumors. Immunohistochemistry of a patient affected with a pituitary adenoma demonstrated the absence of SDHA and SDHB protein, and an increase of prolactin-producing cells. Mutation analysis revealed the *SDHA* mutations, c.725_736del and c.989_990insTA [444]. Another patient also showed a *SDHD* mutation, c.298_301delACTC, along with negative staining of SDHD and SDHB. Both growth hormone receptor and catecholamine were elevated in this patient [115]. Based on these data, it was concluded that hypersecretion of catecholamines and other hormones caused by the mutation in the *SDH* gene resulted in the development of pituitary neoplasm.

The SDH complex has also been reported to play a role in nasopharyngeal carcinoma. Expression of SDHB in recurrent nasopharyngeal carcinoma was significantly lower than that in primary carcinoma. Moreover, a patient with low SDHB expression had a shorter survival period. These indicated that SDH plays a role in the progression of nasopharyngeal carcinoma, and it could be a prognosis marker for this tumor [446].

Furthermore, loss of SDHAF2 has been associated with high epithelial cell transition and lymph node metastasis in lung cancer [447]. *SDHB* silencing with specific siRNA induced cell proliferation, invasion, and migration, but inhibited apoptosis. In contrast, overexpression of SDHB inhibited cell proliferation, invasion, and migration, but promoted apoptosis of ovarian cancer cell lines [448]. Last, low SDH expression in breast cancer was associated with a younger age at diagnosis and low-grade histology [449].

Mechanism of Tumorigenesis Caused by Succinate Dehydrogenase Deficiency

The molecular mechanisms of tumorigenesis in SDH-deficient tumors are still unclear. There are two reasonable models that have been proposed to describe how the loss of SDH function leads to tumor development. These include the cellular response to increased ROS and accumulated succinate.

Previously, the production of ROS within mitochondria has been reported to be influenced by complex I and III. However, recent studies suggested that complex II dysfunction, as observed for those with *SDHx* gene mutations, can lead to the oxidative stress. In a *Caenorhabditis elegans* model, mutation in *MEV-1*, which encodes for the SDHC subunit, as well as *SDHB-1*, which encodes for the SDHB subunit, enhanced superoxide production. Moreover, these mutations were found to result in hypersensitivity to hypoxia, shortened life spans, and impaired complex II activity [450, 451]. In a further study in hamster lung fibroblasts, stable expression of mutant SDHD also increased the steady-state level of ROS, which mediated the genomic instability [452]. Although the direct effects of elevated ROS on tumorigenesis have not been elucidated yet, oxidative stress generally causes DNA, lipid, and protein damage, leading to irregular cellular function that contributes to tumor development. Additionally, it is possible that ROS might promote tumor formation by inducing a pseudohypoxia response. The experimental evidence for this hypothesis was demonstrated in *SDHB*-knockdown pheochromocytoma cells where an increase of ROS production along with the translocation of HIF-1α, a master transcriptional regulator of cellular response to hypoxia, into the nucleus was observed [453].

The second model is based on the hypothesis that the loss of SDH leads to the accumulation of succinate. HIF is a heterodimer comprising HIFα and HIFβ. In normoxic conditions, the protein level of HIF-α is very low because of proteasomal degradation and is associated with ubiquitination by ubiquitin ligase targeted to HIF-α by the VHL protein. Binding of VHL to HIF-α requires hydroxylation of proline residues by prolyl hydroxylase domain (PHD), an oxygen-dependent enzyme that catalyzes the conversion of α-KG to succinate. In hypoxia, PHD activity is decreased, resulting in the reduction of hydroxylation, lack of VHL-mediated ubiquitination, and consequently, accumulation of HIF-α. HIF-α is then translocated into the nucleus and interacts with HIF-β to form an active transcription factor that can upregulate the expression of hypoxic response genes, including genes involved in metabolism, angiogenesis, cell proliferation, and cell survival [454]. Because succinate is a product of PDH, the accumulation of succinate caused by SDH complex dysfunction feedback inhibits the activity of PDH, which in turn leads to downstream signaling similar to what occurs in the hypoxic condition. These observations suggest the development of pseudohypoxia in SDH deficiency. A study supporting this hypothesis was demonstrated in *SDHB*-silenced cell lines. The upregulation of HIF-α was observed in *SDHB*-silenced cells without an increase in ROS level. Moreover, a microarray combined with RT-PCR analysis revealed the overexpression of genes involved in metabolism, cell proliferation, and cell adhesion, some of which are considered as cancer diagnostic markers, and are known for their roles in tumor development and progression [455]. In a further study in SDHB-deficient RCC cell lines, increased HIF-α was observed along with elevated succinate [456].

In addition to pseudohypoxia induction, accumulated succinate is also proposed to causes epigenetic alterations. In Letouze's study, accumulated succinate was observed in tumor samples from *SDH*-mutated PGLs. Moreover, *SDH*-mutated tumors had a low 5-hmC level, abundant histone 3 Lys 27 trimethylation (H3K27me3), and DNA hypermethylation, all changes associated with the downregulation of known tumor suppressor genes [457]. Another study was performed in tumor samples from GIST patients. It was found that 5-hmc staining was absent in 90% of SDH-deficient GISTs, whereas loss of 5-hmC staining was observed in approximately 20% of KIT-or PDGFRA-mutant GISTs [458]. These results were supported by a study in *SDHB* knockdown RCC cell lines that showed a CpG island hypermethylation phenotype along with an elevated succinate level [456].

α-KG-dependent dioxygenases are a large enzyme family, comprising numerous enzymes involved in various cellular biological processes. Among these are the histone demethylases and TET family of DNA hydroxylases, two groups of enzymes involved in transcriptional activation. Histone demethylases remove a methyl group from the lysine residues of histone tails, resulting in open chromatin. TET dioxygenases are responsible for converting 5-mC to 5-hmC, which is required for DNA demethylation. The accumulation of succinate adversely affects the activity of these enzymes [459]. Therefore, it is possible that the succinate accumulation caused by SDH complex dysfunction leads to enzymatic inhibition of multiple α-KG-dependent dioxygenases, contributing to epigenetic alterations. This hypothesis has been confirmed by experimentation in SDH-mutated PGLs. Immunohistochemical analysis demonstrated loss of 5-hmC in SDH-deficient tumor cells. This loss was associated with nuclear exclusion of TET1. Furthermore, increased methylation of H3K9me3 was observed in SDH-deficient tumors, while no changes were seen in H3K4me3 and H3K27me3 [460].

In conclusion, there are several possible mechanisms involved in tumor cell formation. Increased ROS production, pseudohypoxia phenomenon, and epigenetic alterations mediated by the SDH complex defects have been proposed as important underlying mechanisms in tumorigenesis.

Succinate Dehydrogenase and Neurodegenerative Disease

Leigh syndrome is a progressive neurodegenerative disorder that is frequently caused by abnormalities of the mitochondrial respiratory chain, including defects in electron transport chain complex I–V, deficiency in pyruvate dehydrogenase, and mutation in mitochondrial DNA. Deficiencies in complex II are only found in approximately 1–4% cases of Leigh syndrome. Patients with Leigh syndrome typically present with symptoms from an early age and are related to the brain necrotizing. The most common clinical signs of Leigh syndrome are growth retardation, psychomotor delay or regression, muscular hypotonia, cardiomyopathy, respiratory insufficiency, lactic acidosis, and seizures [461].

Leigh syndrome caused by loss of SDH is associated with mutation in the *SDHA* gene. A mutation in the nuclear-encoded flavoprotein subunit gene or *SDHA* was reported in two siblings with complex II deficient Leigh syndrome. Both patients and their parents were homozygous and heterozygous for an Arg554Trp substitution in *SDHA*, respectively. This substitution was observed to decrease the catalytic activity of the SDH complex in a yeast model [462]. SDH-deficient Leigh syndrome was found in a patient who was identified to have compound mutations in the *SDHA* gene [463]. A homozygous Gly555Glu mutation was observed in infants and children who were diagnosed with Leigh syndrome [464, 465]. Another case was found in a 9-year-old German girl. Sequencing of *SDHA* revealed a missense mutation in exon 3 (Ala83Val) and a nonsense mutation in exon 4 (Trp119stop) [466].

Leukoencephalopathy, caused by degeneration of white matter in the brain, has also been associated with mutation in the *SDHx* genes. Alston et al. demonstrated the mutation in the *SDH* genes in two patients. The first patient who presented with leukodystrophy and cardiomyopathy was found to have the *SDHA* compound mutations, p.Thr508Ile and p.Ser509Leu. The second patient who presented with leukodystrophy and hypotonia was found to have the *SDHB* mutation, p.Asp48Val, along with elevated brain succinate. Western blotting confirmed decreased levels of the relevant SDH subunits and impairment of complex II assembly [467]. An Asp48Val mutation in *SDHB* was then reported by Ardissone et al. in one patient who presented with severe hypotonia and psychomotor regression with leukoencephalopathy [468].

Jackson and colleagues reported the first case of leukoencephalopathy with compound *SDHD* gene mutations, including a Glu69Lys missense variant, and leucine insertion at position 164. The patient presented with a lack of SDHD subunit and assembled SDH complex, resulting in significantly decreased complex II activity. Complementation studies of *SDHD* wild-type and mutant were performed in fibroblast cell lines. It was found that only *SDHD* wild-type was able to restore the amount of assembled complex II [469].

Besides mutations in the *SDHx* subunit, the loss of complex II from mutations in an assembly factor gene has been determined in this disease. Ghezzi et al. reported mutations in *SDHAF1* in infantile leukoencephalopathy with defective SDH, including c.169G > C, corresponding to Gly57Arg in the Italian individuals, and c.164G > C, corresponding to Arg55Pro, in the Turkish individuals. Complementation studies in a yeast model were studied by disrupting the *YDR379C-A* gene, the ortholog of *SDHAF1*. The *YDR379C-A* mutant revealed SDH deficiency and failure of oxidative phosphorylation-dependent growth, whereas SDH activity and amount were restored in a mutant with re-expression of the wild-type gene [470]. Other cases of *SDHAF1*-related infantile leukoencephalopathy were observed in five patients affected with accumulation of succinate in their cerebral white matter. Two patients who were siblings carried an Arg55Pro missense mutation that was previously described in Ghezzi's study. Postmortem examination of the elder sister revealed histopathology of encephalomyelopathy with Leigh syndrome. The second sibling carried the missense mutation c.170G > A, corresponding to p.Gly57Glu. Last, one patient carried the nonsense mutation c.22C > T, predicting for premature stop of translation (p.Gln8stop). All observations revealed that their unaffected parents and healthy siblings were heterozygous carriers for the relevant mutation [471].

Taken together, these data confirmed that defects in gene coding for SDH subunits and assembly factors may increase susceptibility to develop neuropathological features. Moreover, these characteristics are genetic in origin and are usually passed from parents to their offspring. The underlying mechanisms of SDH-mediated neurodegeneration are also being investigated. Van Vranken and colleagues characterized the influence of SDH defects due to SDH assembly factor gene knockdown on neurodegeneration using yeast,

Drosophila, and mammalian cells. Lack of SDHAF4, an assembly factor, impaired the interaction and stability of SDHA and SDHB subunits and consequently SDH complex assembly, resulting in reduced SDH activity. It was found that SDHAF4 mutants died significantly earlier than the controls. Additionally, an excess of ROS production and the hallmark phenotypes of mitochondrial dysfunction were also observed. Bang sensitivity, a mitochondrial dysfunction phenotype, increases susceptibility to mechanically induced paralytic seizures. Although mitochondrial defects were observed in a SDHAF4 mutant, loss of SDH did not affect the ATP level [472].

Therefore, based on these results, it is possible that the neuropathological features observed for SDH deficient individuals are driven by increased ROS production and defective mitochondrial redox state balancing.

Succinate Dehydrogenase and Cardiovascular Disease

Dilated cardiomyopathy (DCM), the major cardiomyopathy, is characterized by a dilated ventricular cavity with systolic dysfunction. The clinical symptom of DCM is heart failure, which is often associated with sudden cardiac arrest, resulting in morbidity and mortality. DCM may be a complication of other diseases or may be acquired. The most common form of this disease is idiopathic, while approximately 5% of patients have familial DCM [473]. Previously, mutations in various genes have been associated with DCM, including genes encoding for cardiac structural proteins, signaling proteins, and metabolic components.

In a case report of a 6 week old infant affected with DCM and ventricular noncompaction, a mitochondrial respiratory chain enzyme assay revealed decreased SDH, suggesting a SDH complex deficiency [474]. This observation suggested that SDH defects are a likely cause of cardiomyopathy.

Levitas et al. studied Bedouin patients who presented with DCM. Determination of respiratory chain enzymes in the skeletal and cardiac muscles showed reduced activity of complex II in the muscle (50–60% residual activity) and marked reduction in the cardiac muscle (15–18% residual activity). Direct sequencing also demonstrated a *SDHA* homozygous mutation, c.1664G > A; p.Gly555Glu, which was previously reported in patients with Leigh syndrome. The healthy sibling and unaffected parents were heterozygous for this mutation, indicating autosomal recessive inheritance [475].

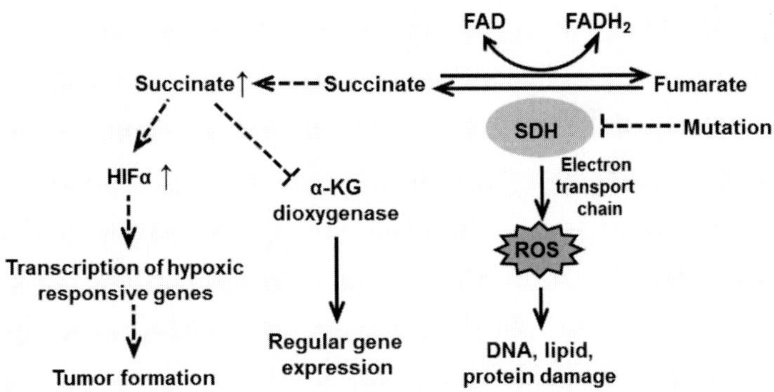

Figure 32. The role of SDH in disease development. SDH produces ROS during transportation of electron from $FADH_2$ into other acceptors. These ROS can react with various biological macromolecules. In the absence of SDK, succinate is accumulated, inhibiting the α-KG dioxygenase, an enzyme functions in epigenetics. Moreover, loss of SDH causes the defect in HIF-1α degradation. HIF-1α signaling activates the expressions of several hypoxic responsive genes, supporting the tumorigenesis. Under oxidative stress and genetic alterations, cellular functions are deregulated.

Further SDH-related cardiomyopathy was observed in a neonate who had left ventricular hypertrophy and died on day one of life. A histochemical assay of a muscle biopsy from this patient revealed a severe reduction of SDH activity that was correlated with the result from a spectrophotometric enzymatic assay. Sequencing of all *SDH* genes identified a novel homozygous c.275A > G, p.Asp92Gly in SDHD, which was heterozygous in his parents. To assess the effect of this variant, an Asp98Gly SDH4 mutant, a change equivalent to the human Asp92Gly SDHD, was constructed in a yeast model. It was found that growth in ethanol-containing media was defective in the mutant, suggesting oxidative growth impairment. Additionally, the oxygen consumption rate of the mutant was 55% less than that of the wild-type [476]. According to the results obtained from the growth experiments and oxygen consumption rate, the SDHD mutation resulted in defective oxidative phosphorylation and may contribute to the pathogenicity of this patient. Contribution of SHD to disease development is demonstrated in Figure 32.

Malate Dehydrogenase

The two major isoforms of malate dehydrogenase (MDH; EC 1.1.1.37) can be differentiated by location, structure and catalysis reaction. MDH2, the mitochondrial form, has a role in the TCA cycle to catalyze the conversion of malate to oxaloacetate (Figure 33). MDH1, the cytosolic form, participates in the malate-aspartate shuttle (MAS), a system that transports reduction equivalents produced during glycolysis across the impermeable mitochondrial membrane. In the cytosol, MDH1 catalyzes the electron transfer from NADH to oxaloacetate to form malate. Malate is then imported into mitochondria via an antiporter following exchange with α-KG. In the matrix, malate is converted into oxaloacetate by MDH2 and electrons are oxidized by NAD^+ to form NADH. This NADH is then used in oxidative phosphorylation for the synthesis of ATP, and oxaloacetate is converted by aspartate aminotransferase into aspartate and then transported into the cytosol by exchange with glutamate. The exported aspartate is then deaminated to oxaloacetate and the cycle is restarted, Figure 34 [477].

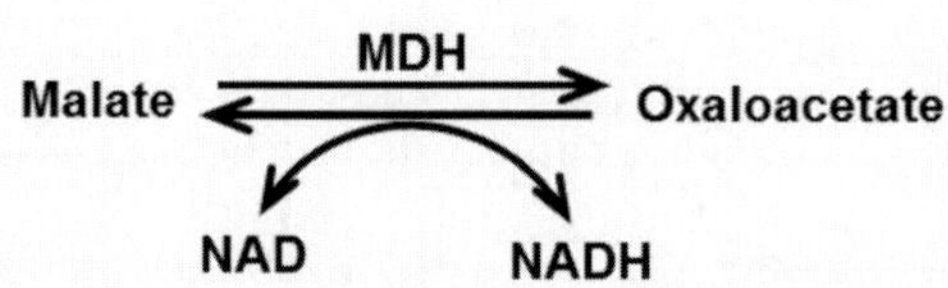

Figure 33. Reaction catalyzed by MDH.

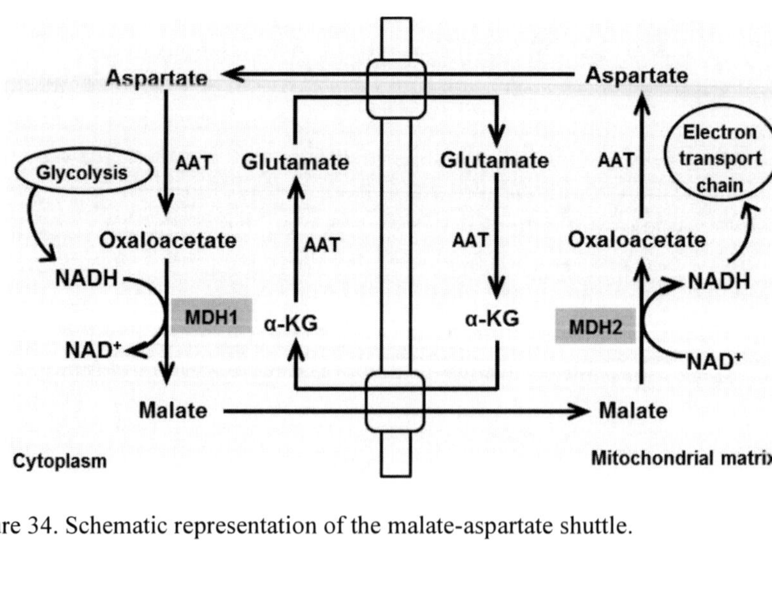

Figure 34. Schematic representation of the malate-aspartate shuttle.

Although there are several components in the MAS, MDH is considered the rate-limiting enzyme in the shuttle [478, 479]. Because MAS mediates the transfer of NADH into mitochondria, changes in MDH can affect cellular energy production, which influences numerous biological processes. Moreover, the imbalance of NAD^+/NADH in the extra- and intra-mitochondria might contribute to the irregular exchange of substances between the mitochondria and cytosol.

Because MDH has a major role in the control of energy production as well as metabolic balance, defects in MDH expression and activity may result in the loss of normal cellular function and homeostasis. Here, a number of MDH-associated diseases, including cancer, cardiovascular, neurologic and bone diseases will be discussed.

Malate Dehydrogenase and Cancer

PGLs are neuroendocrine tumors that arise from sympathetic nervous system paraganglia. Previous studies have reported mutations of several genes, including SDH, a dehydrogenase enzyme that participates in the TCA cycle, are involved in familial PGLs. In a recent study, whole-exome sequencing of tumor DNA obtained from PGL patients demonstrated the presence of many mutations including c.429 + 1G > T, a mutation on the gene encoding *MDH2*. This mutation altered transcripts by introducing a premature stop codon that has the potential to cause transcript elimination via nonsense-mediated decay. This might explain why *MDH2*-mutated patients have low levels of MDH2 activity. *MDH2*-knockdown by shRNA in HeLa cells was used to assess the relationship between MDH2 mutation and PGL development. Malate and fumarate accumulation were observed in the *MDH2*-knockdown cells. Several studies have demonstrated that the accumulation of succinate, caused by SDH mutations, leads to the enzymatic inhibition of multiple α -KG-dependent dioxygenases, and the subsequent impairment of histone demethylation and 5-mC hydroxylation, as well as alterations in DNA methylation. *MDH2*-knockdown cells also exhibited these characteristics as well as genome-wide downregulation [480]. Based on this study, it can be concluded that *MDH2* may act as a tumor suppressor gene, especially as it was reported to be depleted in cancer patients. In addition, MDH2 mutation-mediated metabolic imbalance alters epigenetic modifications that result in cancer cell development.

Interestingly, MDH has also been reported to have a role in the development of chemotherapy-resistant cancer. Doxorubicin, a primary anticancer drug, disturbs DNA synthesis and mitochondrial function resulting in cell apoptosis. However, doxorubicin resistance has been widely reported in various cancers. Doxorubicin-resistant uterine cancer cell lines were generated to investigate the mechanisms of resistance. Mitochondrial proteomic analysis revealed many differentially expressed proteins between drug resistant and control cancers. Of note, MDH2 was overexpressed in highly resistant uterine cancer cells. However, the inhibition of *MDH2* by siRNA did not significantly increase cell apoptosis after drug treatment. Therefore, MDH2 induced doxorubicin-resistant uterine cancer by mechanisms other than those that prevent cell apoptosis [481].

Another MDH-induced chemotherapy resistance study was assessed in docetaxel-resistant prostate cancer. It was found that the *MDH2* gene was overexpressed in nearly 50% of prostate cancer patients. In addition, *MDH2* overexpressing patients exhibited poor outcomes after chemotherapy treatment, suggesting MDH2 regulates the response of prostate cancer to docetaxel treatment. To determine whether MDH2 participates in docetaxel resistance, *MDH2*-knockdown prostate cancer cell lines were generated by *MDH2* shRNA. The *MDH2*-knockdown prostate cancer cells were more sensitive to docetaxel as evidenced by decreased cell viability and increased apoptosis. Docetaxel induces cell apoptosis by damaging microtubules and activating the stress-sensing kinase JNK signaling pathway. The study demonstrated that *MDH2*-knockdown increased docetaxel-induced JNK signaling activation, activated pro-apoptotic proteins, and inactivated anti-apoptotic proteins. These results indicated that MDH2 mediates docetaxel resistance in prostate cancer and that the inhibition of MDH2 might enhance docetaxel-induced JNK signaling activation to increase apoptosis signaling and chemotherapy sensitivity. It was also demonstrated that *MDH2*-knockdown cancer cells had reduced ATP production and mitochondrial NADH. Inadequate NADH production leads to the conversion of oxygen consumed from cellular respiration to ROS. This may be an additional factor that contributes to the increased susceptibility of *MDH2*-knockdown cancer cells to docetaxel [482].

Although the underlying mechanism involved in the development of chemotherapy-resistant cancer caused by the overexpression of MDH2 has not been clearly elucidated, it could be hypothesized that the up-regulation of MDH2 might increase the energy supply for drug-resistant cancers, which will facilitate their survival under drug-induced stress. Moreover, MDH2 might behave as a transcription factor, which can activate anti-apoptotic and inhibit pro-apoptotic signaling pathways. Nevertheless, these studies provide evidence that MDH2 might be a potential marker and therapeutic target for chemotherapy-resistant cancer.

Malate Dehydrogenase and Cardiovascular Disease

The role of MDH in heart diseases such as chronic heart failure (CHF) has been demonstrated. A swine model of chronic progressive heart failure was produced by placing an ameroid constrictor on the left anterior descending coronary artery. After evaluation of the CHF model, swine myocardium and plasma were subjected to proteomic and metabolomic analysis, respectively. Decreased MDH and energy metabolites were observed. This finding revealed that MDH-mediated energy metabolism may be a key element in the pathogenesis of CHF. As a result of the down-regulation of MDH, the available energy is insufficient for cardiac contractility and relaxation resulting in the occurrence of CHF clinical signs, including chest tightness, palpitation, laziness, and other symptoms [478].

Another study also demonstrated the relevance of MDH2 and doxorubicin-induced cardiomyopathy. Although doxorubicin is an effective chemotherapy used in cancer treatment, it was recently reported to be cardiotoxic. To identify a compound that might inhibit this adverse effect, zebrafish and mouse models of doxorubicin-induced cardiomyopathy were established. Visnagin was shown to prevent doxorubicin toxicity as evidenced by decreased cardiomyocyte apoptosis and improved cardiac function after co-treatment with doxorubicin. Competitive pull-down assay and mass spectrometry analysis revealed visnagin bound to MDH2, indicating this was the cardioprotective target of visnagin. To examine how MDH2 inhibition mediates the cardioprotective effect of visnagin, MDH2 and MAS inhibitors were administrated and doxorubicin-induced cardiotoxicity was determined. Increased cell viability was observed by either MDH2 or MAS inhibition [483]. In summary, MDH2 is involved in the development of doxorubicin-induced cardiomyopathy; however, further studies are required to elucidate mechanism in greater detail. Remarkably, the cardioprotective effect of MDH2 inhibition provides evidence that MDH2 might be a potential therapeutic target in the combination treatment with doxorubicin, as well as other drug-induced cardiac injuries.

Malate Dehydrogenase and Neurodegenerative Disease

Human G93A-superoxide dismutase-1 mutation (G93AhSOD1) causes amyotrophic lateral sclerosis, a neurodegenerative disorder in humans. Recently, it was reported that G93AhSOD1 can interact with MDH1, interfering in the full enzymatic activity of MDH1. To investigate this toxic function, NSC-34, a mouse motor neuron cell line, was co-transfected with G93AhSOD1 and MDH1. ATP levels in G93AhSOD1 transfected cells were lower than in wild-type hSOD1 cells. Moreover, cell survival under rotenone toxicity and low glucose challenge were also reduced in G93AhSOD1 cells. However, blocking the interaction of G93AhSOD1 with MDH1 by a MDH1-decoy peptide recovered ATP levels and survival of NSC-34 cells [484]. Based on these results, it was concluded that MDH1 is essential for ATP production in neurons, and the blockade of MDH1 activity by G93AhSOD1 causes neuron weakness and reduces neuron tolerance under toxic and metabolic stresses. These factors can in turn lead to the degeneration of motor neurons, ultimately resulting in muscle weakness and atrophy, loss of movement control, and the induction of psychological disorders.

The involvement of MDH2 in neurologic diseases has been suggested as it was elevated in the brains of AD patients that succumbed to the disease. Increased MDH2 was assumed to be caused by oxidative stress, an early and pervasive event in AD. A study of HT22, a mouse hippocampal cell line, revealed that H_2O_2 increased *MDH2* mRNA expression and MDH activity, indicating elevated MDH2 in patient brains may be a response to oxidative stress. Furthermore, it was reported that oxidative stress elevated MDH2 by the down-regulation of miR-743a, a negative regulator of *MDH2* expression. Although the exact function of elevated MDH2 in AD is unknown, it was hypothesized that it might be related to its function in the MAS to maintain a balanced redox state under NADH-reduced oxidative stress. Another possibility is that MDH provides NAD^+ to KGDH, a key TCA cycle enzyme, and other dehydrogenases [485].

Malate Dehydrogenase and Bone Disease

A recent study investigated the role of MDH1 in osteoclastogenesis. Increased *MDH1* mRNA expression and protein levels were observed during osteoclast maturation induced with a RANKL. To assess whether MDH1 participated in osteoclastogenesis, *MDH1*-knockdown bone marrow cells were constructed by infection with shRNA. A decreased formation of mature osteoclasts and the expression of differentiation-related genes were observed along with reduced ATP. These results correlated with previous studies that energy insufficiency caused by MDH deficiency caused cell dysfunction. To determine how reduced ATP modulates osteoclastogenesis, the expressions of signaling proteins were measured. AMPK, an intracellular energy sensor, was increased, while the expression levels of c-Fos and NFATc1, critical transcription factors in osteoclast differentiation, were decreased. These results suggested that *MDH1*-knockdown mediated ATP depletion and activated the AMPK signaling pathway, which in turn downregulated c-Fos and NFATc1 levels [486]. Taken together, these data suggest that MDH1 regulates osteoclast energy. The reduction of ATP caused by defects in MDH1 is detected by AMPK, resulting in the inhibition of transcription factors for osteoclastogenesis. Thus, abnormalities of osteoclast activation and differentiation are involved in bone-related diseases such as osteopetrosis and periodontitis.

Malate Dehydrogenase and the Stress Response

Upon the occurrence of stress, MDH directly modifies macromolecules, including nucleic acids, proteins, carbohydrates, and lipids. These events extend to cellular dysfunction and pathology; however, there are other fundamental processes that have a role in the control of stress and maintenance of cellular homeostasis.

MDH1 has a role in metabolic stress by binding to p53, a transcription factor that controls cell-cycle arrest and cell death, and this interaction is preferred under conditions of glucose deprivation [487]. In addition, increased MDH1 was observed under glucose-deprived conditions. This correlation led to the hypothesis that MDH1 may be involved in the response of p53 to metabolic stress. Furthermore, glucose depletion caused a change in the NAD^+/NADH ratio that may induce MDH1-p53 interactions. The binding of MDH1 to the p53 promoter reduced ubiquitination and increased activation of p53, resulting in cell apoptosis. Thus, this study provided evidence that MDH1 regulates p53 transcription activity in response to metabolic stress [487].

Aging, one of many cellular responses to stress, is associated with MDH1. A study of human dermal fibroblasts revealed a reduction of the NAD^+/NADH ratio in *MDH1*-knockdown cells. Moreover, the loss of MDH1 induced cellular senescence. To assess whether decreased NAD^+/NADH leads to cellular senescence, the expressions of signaling proteins were determined. Increased AMPK and phosphorylated p53 as well as reduced SIRT1, an anti-senescence protein, were reported. These results suggested that a decreased NAD^+/NADH ratio caused by *MDH1*-knockdown triggered the activation of AMPK that stimulates senescent transcription factors and inhibits anti-senescence proteins [488]. Taken together, these data suggest that MDH might function as a signaling effector to control the cellular events in various stress conditions. The role of MDH in development of diseases is present in Figure 35.

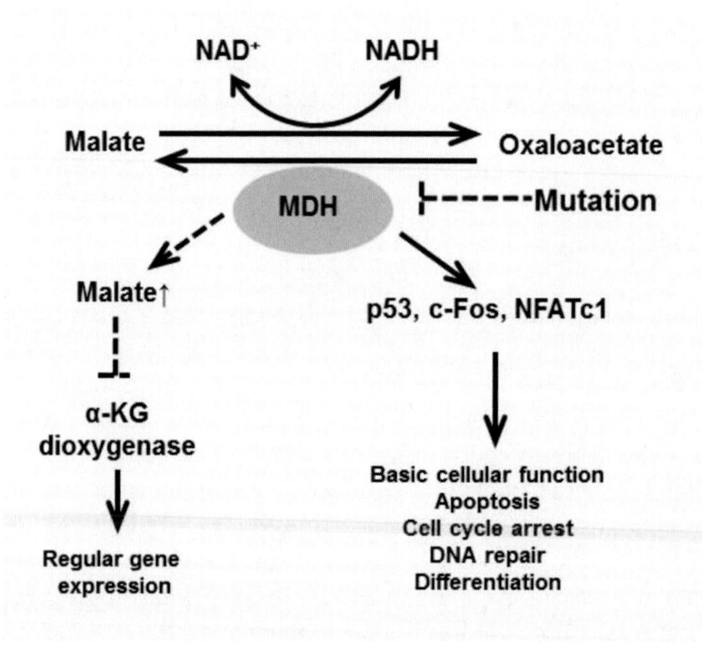

Figure 35. The role of MDH in disease development. Normally, MDH plays role in the controlling of cellular normal function. Defect in MDH can lead to the irregular function, including loss of differentiation of immature cells. These irregular functions caused by MDH deficiency are driven by inactivation of signaling molecules and epigenetics alterations.

CONCLUSION

Alteration in glucose metabolism cuased by defects in dehydrogenases contributes to various disease conditions, including cancer, neurodegenerative, cardiovascular and liver diseases, diabetes and obesity. In cancer, a functionally impaired mitochondrial metabolism is crucial for tumorigenesis. Aerobic glycolysis or the Warburg effect provides a selective advantage for tumor cells to acquire increased growh, proliferation, migration, survival and invasion and to become more resistant to apoptosis. As a consequence, several dehydrogenases are upregulated in cencer.

Dehydrogenases also constitute an integral part in many neurodegenerative disorders. In addition to provide energy metabolism, they are essential for synthesis and degradation of neurotransmitters. Furthermore, they are required to maintain normal function and homeostatis of neural cells.

Though increasing evidence indicates the contribution of dehydrogenases in diseases, the molecular mechanisms underlying the role of dehydrogenases in etiology of many diseases are still unclear. Therefore, further investigations are required. Regardless of the mechanisms, dehydrogenases represent potential targets for diagnosis and treatment of several diseases.

REFERENCES

[1] Nicholls C, Li H, Liu JP (2012) GAPDH: a common enzyme with uncommon functions. *Clin Exp Pharmacol Physiol* 39: 674-679.

[2] Zhang JY, Zhang F, Hong CQ, Giuliano AE, Cui XJ, et al. (2015) Critical protein GAPDH and its regulatory mechanisms in cancer cells. *Cancer Biol Med* 12: 10-22.

[3] Liberti MV, Locasale JW (2016) The Warburg Effect: How does it benefit cancer cells? *Trends Biochem Sci* 41: 211-218.

[4] Hao L, Zhou X, Liu S, Sun M, Song Y, et al. (2015) Elevated GAPDH expression is associated with the proliferation and invasion of lung and esophageal squamous cell carcinomas. *Proteomics* 15: 3087-3100.

[5] Harada N, Yasunaga R, Higashimura Y, Yamaji R, Fujimoto K, et al. (2007) Glyceraldehyde-3-phosphate dehydrogenase enhances transcriptional activity of androgen receptor in prostate cancer cells. *J Biol Chem* 282: 22651-22661.

[6] Mikuriya K, Kuramitsu Y, Ryozawa S, Fujimoto M, Mori S, et al. (2007) Expression of glycolytic enzymes is increased in pancreatic cancerous tissues as evidenced by proteomic profiling by two-dimensional electrophoresis and liquid chromatography-mass spectrometry/mass spectrometry. *Int J Oncol* 30: 849-855.

[7] Ganapathy-Kanniappan S, Kunjithapatham R, Geschwind JF (2012) Glyceraldehyde-3-phosphate dehydrogenase: a promising target for molecular therapy in hepatocellular carcinoma. *Oncotarget* 3: 940-953.

[8] Bazzocco S, Dopeso H, Carton-Garcia F, Macaya I, Andretta E, et al. (2015) Highly expressed genes in rapidly proliferating tumor cells as new targets for colorectal cancer treatment. *Clin Cancer Res* 21: 3695-3704.

[9] Wang D, Moothart DR, Lowy DR, Qian X (2013) The expression of glyceraldehyde-3-phosphate dehydrogenase associated cell cycle (GACC) genes correlates with cancer stage and poor survival in patients with solid tumors. *PLoS One* 8: e61262.

[10] Puzone R, Savarino G, Salvi S, Dal Bello MG, Barletta G, et al. (2013) Glyceraldehyde-3-phosphate dehydrogenase gene over expression correlates with poor prognosis in non small cell lung cancer patients. *Mol Cancer* 12: 97.

[11] Colell A, Ricci J-E, Tait S, Milasta S, Maurer U, et al. (2007) GAPDH and autophagy preserve survival after apoptotic cytochrome c release in the absence of caspase activation. *Cell* 129: 983-997.

[12] Carujo S, Estanyol JM, Ejarque A, Agell N, Bachs O, et al. (2006) Glyceraldehyde 3-phosphate dehydrogenase is a SET-binding protein and regulates cyclin B-cdk1 activity. *Oncogene* 25: 4033-4042.

[13] Zhou Y, Yi X, Stoffer JB, Bonafe N, Gilmore-Hebert M, et al. (2008) The multifunctional protein glyceraldehyde-3-phosphate dehydrogenase is both regulated and controls colony-stimulating factor-1 messenger RNA stability in ovarian cancer. *Mol Cancer Res* 6: 1375-1384.

[14] Demarse NA, Ponnusamy S, Spicer EK, Apohan E, Baatz JE, et al. (2009) Direct binding of glyceraldehyde 3-phosphate dehydrogenase to telomeric DNA protects telomeres against chemotherapy-induced rapid degradation. *J Mol Biol* 394: 789-803.

[15] Azam S, Jouvet N, Jilani A, Vongsamphanh R, Yang X, et al. (2008) Human glyceraldehyde-3-phosphate dehydrogenase plays a direct role in reactivating oxidized forms of the DNA repair enzyme APE1. *J Biol Chem* 283: 30632-30641.

[16] Tatton NA (2000) Increased caspase 3 and Bax immunoreactivity accompany nuclear *GAPDH* translocation and neuronal apoptosis in Parkinson's disease. *Exp Neurol* 166: 29-43.

[17] Huang J, Xiong N, Chen C, Xiong J, Jia M, et al. (2011) Glyceraldehyde-3-phosphate dehydrogenase: activity inhibition and

protein overexpression in rotenone models for Parkinson's disease. *Neuroscience* 192: 598-608.
[18] Sunaga K, Takahashi H, Chuang DM, Ishitani R (1995) Glyceraldehyde-3-phosphate dehydrogenase is over-expressed during apoptotic death of neuronal cultures and is recognized by a monoclonal antibody against amyloid plaques from Alzheimer's brain. *Neurosci Lett* 200: 133-136.
[19] Mazzola JL, Sirover MA (2001) Reduction of glyceraldehyde-3-phosphate dehydrogenase activity in Alzheimer's disease and in Huntington's disease fibroblasts. *J Neurochem* 76: 442-449.
[20] Mazzola JL, Sirover MA (2003) Subcellular alteration of glyceraldehyde-3-phosphate dehydrogenase in Alzheimer's disease fibroblasts. *J Neurosci Res* 71: 279-285.
[21] Shalova IN, Cechalova K, Rehakova Z, Dimitrova P, Ognibene E, et al. (2007) Decrease of dehydrogenase activity of cerebral glyceraldehyde-3-phosphate dehydrogenase in different animal models of Alzheimer's disease. *Biochim Biophys Acta* 1770: 826-832.
[22] Rocha NP, Ribeiro FM, Furr-Stimming E, Teixeira AL (2016) Neuroimmunology of Huntington's Disease: Revisiting Evidence from Human Studies. *Mediators Inflamm* 2016: 8653132.
[23] Burke JR, Enghild JJ, Martin ME, Jou YS, Myers RM, et al. (1996) Huntingtin and DRPLA proteins selectively interact with the enzyme GAPDH. *Nat Med* 2: 347-350.
[24] Mazzola JL, Sirover MA (2002) Alteration of nuclear glyceraldehyde-3-phosphate dehydrogenase structure in Huntington's disease fibroblasts. *Brain Res Mol Brain Res* 100: 95-101.
[25] Senatorov VV, Charles V, Reddy PH, Tagle DA, Chuang DM (2003) Overexpression and nuclear accumulation of glyceraldehyde-3-phosphate dehydrogenase in a transgenic mouse model of Huntington's disease. *Mol Cell Neurosci* 22: 285-297.
[26] Bae BI, Hara MR, Cascio MB, Wellington CL, Hayden MR, et al. (2006) Mutant huntingtin: nuclear translocation and cytotoxicity mediated by GAPDH. *Proc Natl Acad Sci U S A* 103: 3405-3409.
[27] Beisswenger PJ, Howell SK, Smith K, Szwergold BS (2003) Glyceraldehyde-3-phosphate dehydrogenase activity as an independent modifier of methylglyoxal levels in diabetes. *Biochim Biophys Acta* 1637: 98-106.
[28] Kanwar M, Kowluru RA (2009) Role of glyceraldehyde 3-phosphate dehydrogenase in the development and progression of diabetic retinopathy. *Diabetes* 58: 227-234.

[29] Madsen-Bouterse S, Mohammad G, Kowluru RA (2010) Glyceraldehyde-3-phosphate dehydrogenase in retinal microvasculature: implications for the development and progression of diabetic retinopathy. *Invest Ophthalmol Vis Sci* 51: 1765-1772.

[30] Blatnik M, Thorpe SR, Baynes JW (2008) Succination of proteins by fumarate: mechanism of inactivation of glyceraldehyde-3-phosphate dehydrogenase in diabetes. *Ann N Y Acad Sci* 1126: 272-275.

[31] Cordes W (1926) Experiences with plasmochin in malaria (preliminary reports). Boston. 66 p.

[32] Toncheva D, Tzoneva M (1985) Prenatal selection and fetal development disturbances occurring in carriers of G6PD deficiency. *Hum Genet* 69: 88.

[33] Sodeinde O (1992) Glucose-6-phosphate dehydrogenase deficiency. *Baillieres Clin Haematol* 5: 367-382.

[34] Beutler E (1994) G6PD deficiency. *Blood* 84: 3613-3636.

[35] Iwai K, Hirono A, Matsuoka H, Kawamoto F, Horie T, et al. (2001) Distribution of glucose-6-phosphate dehydrogenase mutations in Southeast Asia. *Hum Genet* 108: 445-449.

[36] Tishkoff SA, Varkonyi R, Cahinhinan N, Abbes S, Argyropoulos G, et al. (2001) Haplotype diversity and linkage disequilibrium at human G6PD: recent origin of alleles that confer malarial resistance. *Science* 293: 455-462.

[37] Nuchprayoon I, Sanpavat S, Nuchprayoon S (2002) Glucose-6-phosphate dehydrogenase (G6PD) mutations in Thailand: G6PD Viangchan (871G>A) is the most common deficiency variant in the Thai population. *Hum Mutat* 19: 185.

[38] Sabeti PC, Reich DE, Higgins JM, Levine HZ, Richter DJ, et al. (2002) Detecting recent positive selection in the human genome from haplotype structure. *Nature* 419: 832-837.

[39] Louicharoen C, Nuchprayoon I (2005) G6PD Viangchan (871G>A) is the most common G6PD-deficient variant in the Cambodian population. *J Hum Genet* 50: 448-452.

[40] Guindo A, Fairhurst RM, Doumbo OK, Wellems TE, Diallo DA (2007) X-linked G6PD deficiency protects hemizygous males but not heterozygous females against severe malaria. *PLoS Med* 4: e66.

[41] Ruwende C, Khoo SC, Snow RW, Yates SN, Kwiatkowski D, et al. (1995) Natural selection of hemi- and heterozygotes for G6PD deficiency in Africa by resistance to severe malaria. *Nature* 376: 246-249.

[42] Roth EF, Jr., Raventos-Suarez C, Rinaldi A, Nagel RL (1983) Glucose-6-phosphate dehydrogenase deficiency inhibits *in vitro* growth of *Plasmodium falciparum*. *Proc Natl Acad Sci U S A* 80: 298-299.

[43] Louicharoen C, Patin E, Paul R, Nuchprayoon I, Witoonpanich B, et al. (2009) Positively selected G6PD-Mahidol mutation reduces *Plasmodium vivax* density in Southeast Asians. *Science* 326: 1546-1549.

[44] Xu W, Westwood B, Bartsocas CS, Malcorra-Azpiazu JJ, Indrak K, et al. (1995) Glucose-6 phosphate dehydrogenase mutations and haplotypes in various ethnic groups. *Blood* 85: 257-263.

[45] Kotaka M, Gover S, Vandeputte-Rutten L, Au SW, Lam VM, et al. (2005) Structural studies of glucose-6-phosphate and $NADP^+$ binding to human glucose-6-phosphate dehydrogenase. *Acta Crystallogr D Biol Crystallogr* 61: 495-504.

[46] Minucci A, Moradkhani K, Hwang MJ, Zuppi C, Giardina B, et al. (2012) Glucose-6-phosphate dehydrogenase (G6PD) mutations database: review of the "old" and update of the new mutations. *Blood Cells Mol Dis* 48: 154-165.

[47] Au SW, Gover S, Lam VM, Adams MJ (2000) Human glucose-6-phosphate dehydrogenase: the crystal structure reveals a structural $NADP(^+)$ molecule and provides insights into enzyme deficiency. *Structure* 8: 293-303.

[48] Boonyuen U, Chamchoy K, Swangsri T, Saralamba N, Day NP, et al. (2016) Detailed functional analysis of two clinical glucose-6-phosphate dehydrogenase (G6PD) variants, $G6PD_{Viangchan}$ and $G6PD_{Viangchan+Mahidol}$: Decreased stability and catalytic efficiency contribute to the clinical phenotype. *Mol Genet Metab* 118: 84-91.

[49] Wang XT, Lam VM, Engel PC (2005) Marked decrease in specific activity contributes to disease phenotype in two human glucose 6-phosphate dehydrogenase mutants, G6PD(Union) and G6PD(Andalus). *Hum Mutat* 26: 284.

[50] Wang XT, Lam VM, Engel PC (2006) Functional properties of two mutants of human glucose 6-phosphate dehydrogenase, R393G and R393H, corresponding to the clinical variants G6PD Wisconsin and Nashville. *Biochim Biophys Acta* 1762: 767-774.

[51] Bouanga JC, Mouele R, Prehu C, Wajcman H, Feingold J, et al. (1998) Glucose-6-phosphate dehydrogenase deficiency and homozygous sickle cell disease in Congo. *Hum Hered* 48: 192-197.

[52] Lesesve JF, Perrin J (2012) Association of homozygous sickle cell anemia and glucose-6-phosphate dehydrogenase deficiency. *Eur J Haematol* 88: 370.

[53] Kaplan M, Hammerman C (2002) Glucose-6-phosphate dehydrogenase deficiency: a potential source of severe neonatal hyperbilirubinaemia and kernicterus. *Semin Neonatol* 7: 121-128.

[54] Ho HY, Cheng ML, Weng SF, Chang L, Yeh TT, et al. (2008) Glucose-6-phosphate dehydrogenase deficiency enhances enterovirus 71 infection. *J Gen Virol* 89: 2080-2089.

[55] Fujino G, Noguchi T, Takeda K, Ichijo H (2006) Thioredoxin and protein kinases in redox signaling. *Semin Cancer Biol* 16: 427-435.

[56] Wong WR, Chen YY, Yang SM, Chen YL, Horng JT (2005) Phosphorylation of PI3K/Akt and MAPK/ERK in an early entry step of enterovirus 71. *Life Sci* 78: 82-90.

[57] Wu YH, Tseng CP, Cheng ML, Ho HY, Shih SR, et al. (2008) Glucose-6-phosphate dehydrogenase deficiency enhances human coronavirus 229E infection. *J Infect Dis* 197: 812-816.

[58] Chao YC, Huang CS, Lee CN, Chang SY, King CC, et al. (2008) Higher infection of dengue virus serotype 2 in human monocytes of patients with G6PD deficiency. *PLoS One* 3: e1557.

[59] Wu YH, Chiu DT, Lin HR, Tang HY, Cheng ML, et al. (2015) Glucose-6-phosphate dehydrogenase enhances antiviral response through downregulation of NADPH sensor HSCARG and upregulation of NF-kB signaling. *Viruses* 7: 6689-6706.

[60] Hong X, Song R, Song H, Zheng T, Wang J, et al. (2014) PTEN antagonises Tcl1/hnRNPK-mediated G6PD pre-mRNA splicing which contributes to hepatocarcinogenesis. *Gut* 63: 1635-1647.

[61] Kong DH, Li S, Du ZX, Liu C, Liu BQ, et al. (2016) BAG3 elevation inhibits cell proliferation via direct interaction with G6PD in hepatocellular carcinomas. *Oncotarget* 7: 700-711.

[62] Du W, Jiang P, Mancuso A, Stonestrom A, Brewer MD, et al. (2013) TAp73 enhances the pentose phosphate pathway and supports cell proliferation. *Nat Cell Biol* 15: 991-1000.

[63] Jiang P, Du W, Yang X (2013) A critical role of glucose-6-phosphate dehydrogenase in TAp73-mediated cell proliferation. *Cell Cycle* 12: 3720-3726.

[64] Hu T, Li YS, Chen B, Chang YF, Liu GC, et al. (2015) Elevated glucose-6-phosphate dehydrogenase expression in the cervical cancer

cases is associated with the cancerigenic event of high-risk human papillomaviruses. *Exp Biol Med (Maywood)* 240: 1287-1297.

[65] Manoharan S, Kolanjiappan K, Kayalvizhi M (2004) Enhanced lipid peroxidation and impaired enzymic antioxidant activities in the erythrocytes of patients with cervical carcinoma. *Cell Mol Biol Lett* 9: 699-707.

[66] Mobley JA, Brueggemeier RW (2004) Estrogen receptor-mediated regulation of oxidative stress and DNA damage in breast cancer. *Carcinogenesis* 25: 3-9.

[67] Pu H, Zhang Q, Zhao C, Shi L, Wang Y, et al. (2015) Overexpression of G6PD is associated with high risks of recurrent metastasis and poor progression-free survival in primary breast carcinoma. *World J Surg Oncol* 13: 323.

[68] Dong T, Kang X, Liu Z, Zhao S, Ma W, et al. (2016) Altered glycometabolism affects both clinical features and prognosis of triple-negative and neoadjuvant chemotherapy-treated breast cancer. *Tumour Biol* 37: 8159-8168.

[69] Yi H, Zheng X, Song J, Shen R, Su Y, et al. (2015) Exosomes mediated pentose phosphate pathway in ovarian cancer metastasis: a proteomics analysis. *Int J Clin Exp Pathol* 8: 15719-15728.

[70] Catanzaro D, Gaude E, Orso G, Giordano C, Guzzo G, et al. (2015) Inhibition of glucose-6-phosphate dehydrogenase sensitizes cisplatin-resistant cells to death. *Oncotarget* 6: 30102-30114.

[71] Wang X, Liu H, Zhang X, Li X, Gu H, et al. (2016) G6PD downregulation triggered growth inhibition and induced apoptosis by regulating STAT3 signaling pathway in esophageal squamous cell carcinoma. *Tumour Biol* 37: 781-789.

[72] Li D, Zhu Y, Tang Q, Lu H, Li H, et al. (2009) A new G6PD knockdown tumor-cell line with reduced proliferation and increased susceptibility to oxidative stress. *Cancer Biother Radiopharm* 24: 81-90.

[73] Hu T, Zhang C, Tang Q, Su Y, Li B, et al. (2013) Variant G6PD levels promote tumor cell proliferation or apoptosis via the STAT3/5 pathway in the human melanoma xenograft mouse model. *BMC Cancer* 13: 251.

[74] Cai T, Kuang Y, Zhang C, Zhang Z, Chen L, et al. (2015) Glucose 6-phosphate dehydrogenase and NADPH oxidase 4 control STAT3 activity in melanoma cells through a pathway involving reactive oxygen species, c-SRC and SHP2. *Am J Cancer Res* 5: 1610-1620.

[75] Tome ME, Johnson DB, Samulitis BK, Dorr RT, Briehl MM (2006) Glucose 6-phosphate dehydrogenase overexpression models glucose

deprivation and sensitizes lymphoma cells to apoptosis. *Antioxid Redox Signal* 8: 1315-1327.

[76] Lucarelli G, Galleggiante V, Rutigliano M, Sanguedolce F, Cagiano S, et al. (2015) Metabolomic profile of glycolysis and the pentose phosphate pathway identifies the central role of glucose-6-phosphate dehydrogenase in clear cell-renal cell carcinoma. *Oncotarget* 6: 13371-13386.

[77] Vizan P, Alcarraz-Vizan G, Diaz-Moralli S, Solovjeva ON, Frederiks WM, et al. (2009) Modulation of pentose phosphate pathway during cell cycle progression in human colon adenocarcinoma cell line HT29. *Int J Cancer* 124: 2789-2796.

[78] Jiang P, Du W, Wang X, Mancuso A, Gao X, et al. (2011) p53 regulates biosynthesis through direct inactivation of glucose-6-phosphate dehydrogenase. *Nat Cell Biol* 13: 310-316.

[79] Polimeni M, Voena C, Kopecka J, Riganti C, Pescarmona G, et al. (2011) Modulation of doxorubicin resistance by the glucose-6-phosphate dehydrogenase activity. *Biochem J* 439: 141-149.

[80] Zampella EJ, Bradley EL, Jr., Pretlow TG, 2nd (1982) Glucose-6-phosphate dehydrogenase: a possible clinical indicator for prostatic carcinoma. *Cancer* 49: 384-387.

[81] Duvel K, Yecies JL, Menon S, Raman P, Lipovsky AI, et al. (2010) Activation of a metabolic gene regulatory network downstream of mTOR complex 1. *Mol Cell* 39: 171-183.

[82] Tsouko E, Khan AS, White MA, Han JJ, Shi Y, et al. (2014) Regulation of the pentose phosphate pathway by an androgen receptor-mTOR-mediated mechanism and its role in prostate cancer cell growth. *Oncogenesis* 3: e103.

[83] Rao X, Duan X, Mao W, Li X, Li Z, et al. (2015) O-GlcNAcylation of G6PD promotes the pentose phosphate pathway and tumor growth. *Nat Commun* 6: 8468.

[84] Wang YP, Zhou LS, Zhao YZ, Wang SW, Chen LL, et al. (2014) Regulation of G6PD acetylation by SIRT2 and KAT9 modulates NADPH homeostasis and cell survival during oxidative stress. *EMBO J* 33: 1304-1320.

[85] Lou MF (2000) Thiol regulation in the lens. *J Ocul Pharmacol Ther* 16: 137-148.

[86] Orzalesi N, Sorcinelli R, Guiso G (1981) Increased incidence of cataracts in male subjects deficient in glucose-6-phosphate dehydrogenase. *Arch Ophthalmol* 99: 69-70.

[87] Moro F, Gorgone G, Li Volti S, Cavallaro N, Faro S, et al. (1985) Glucose-6-phosphate dehydrogenase deficiency and incidence of cataract in Sicily. *Ophthalmic Paediatr Genet* 5: 197-200.

[88] Orzalesi N, Fogagnolo P, Rossetti L (2008) Intraocular pressure fluctuations in glaucoma. *Arch Ophthalmol* 126: 745; author reply 745.

[89] Hayashi R, Hayashi S, Arai K, Chikuda M, Obara Y (2012) Effects of antioxidant supplementation on mRNA expression of glucose-6-phosphate dehydrogenase, beta-actin and 18S rRNA in the anterior capsule of the lens in cataract patients. *Exp Eye Res* 96: 48-54.

[90] Dauer W, Przedborski S (2003) Parkinson's disease: mechanisms and models. *Neuron* 39: 889-909.

[91] Dawson TM, Dawson VL (2003) Rare genetic mutations shed light on the pathogenesis of Parkinson disease. *J Clin Invest* 111: 145-151.

[92] Alam ZI, Jenner A, Daniel SE, Lees AJ, Cairns N, et al. (1997) Oxidative DNA damage in the parkinsonian brain: an apparent selective increase in 8-hydroxyguanine levels in substantia nigra. *J Neurochem* 69: 1196-1203.

[93] Migliore L, Petrozzi L, Lucetti C, Gambaccini G, Bernardini S, et al. (2002) Oxidative damage and cytogenetic analysis in leukocytes of Parkinson's disease patients. *Neurology* 58: 1809-1815.

[94] Abraham S, Soundararajan CC, Vivekanandhan S, Behari M (2005) Erythrocyte antioxidant enzymes in Parkinson's disease. *Indian J Med Res* 121: 111-115.

[95] Garcia-Nogales P, Almeida A, Bolanos JP (2003) Peroxynitrite protects neurons against nitric oxide-mediated apoptosis. A key role for glucose-6-phosphate dehydrogenase activity in neuroprotection. *J Biol Chem* 278: 864-874.

[96] Mejias R, Villadiego J, Pintado CO, Vime PJ, Gao L, et al. (2006) Neuroprotection by transgenic expression of glucose-6-phosphate dehydrogenase in dopaminergic nigrostriatal neurons of mice. *J Neurosci* 26: 4500-4508.

[97] Jeng W, Loniewska MM, Wells PG (2013) Brain glucose-6-phosphate dehydrogenase protects against endogenous oxidative DNA damage and neurodegeneration in aged mice. *ACS Chem Neurosci* 4: 1123-1132.

[98] Russell RL, Siedlak SL, Raina AK, Bautista JM, Smith MA, et al. (1999) Increased neuronal glucose-6-phosphate dehydrogenase and sulfhydryl levels indicate reductive compensation to oxidative stress in Alzheimer disease. *Arch Biochem Biophys* 370: 236-239.

[99] Palmer CV, Adams SW, Bourgeois M, Durrant J, Rossi M (1999) Reduction in caregiver-identified problem behaviors in patients with Alzheimer disease post-hearing-aid fitting. *J Speech Lang Hear Res* 42: 312-328.

[100] Wan GH, Tsai SC, Chiu DT (2002) Decreased blood activity of glucose-6-phosphate dehydrogenase associates with increased risk for diabetes mellitus. *Endocrine* 19: 191-195.

[101] Heymann AD, Cohen Y, Chodick G (2012) Glucose-6-phosphate dehydrogenase deficiency and type 2 diabetes. *Diabetes Care* 35: e58.

[102] Park J, Rho HK, Kim KH, Choe SS, Lee YS, et al. (2005) Overexpression of glucose-6-phosphate dehydrogenase is associated with lipid dysregulation and insulin resistance in obesity. *Mol Cell Bio*l 25: 5146-5157.

[103] Wang F, Zhao Y, Niu Y, Wang C, Wang M, et al. (2012) Activated glucose-6-phosphate dehydrogenase is associated with insulin resistance by upregulating pentose and pentosidine in diet-induced obesity of rats. *Horm Metab Res* 44: 938-942.

[104] Ham M, Choe SS, Shin KC, Choi G, Kim JW, et al. (2016) Glucose-6-phosphate dehydrogenase deficiency improves insulin resistance with reduced adipose tissue inflammation in obesity. *Diabetes* 65: 2624-2638.

[105] Sankhla M, Mathur K, Rathor JS (2012) Is there any role of glucose-6-phosphate dehydrogenase in obesity induced metabolic disorder.pdf. *Health* 4: 1530-1536.

[106] Xu Y, Osborne BW, Stanton RC (2005) Diabetes causes inhibition of glucose-6-phosphate dehydrogenase via activation of PKA, which contributes to oxidative stress in rat kidney cortex. *Am J Physiol Renal Physiol* 289: F1040-1047.

[107] McDermott BM, Flatt PR, Strain JJ (1994) Effects of copper deficiency and experimental diabetes on tissue antioxidant enzyme levels in rats. *Ann Nutr Metab* 38: 263-269.

[108] Ulusu NN, Sahilli M, Avci A, Canbolat O, Ozansoy G, et al. (2003) Pentose phosphate pathway, glutathione-dependent enzymes and antioxidant defense during oxidative stress in diabetic rodent brain and peripheral organs: effects of stobadine and vitamin E. *Neurochem Res* 28: 815-823.

[109] Muggeo M, Moghetti P, Querena M, Cacciatori V, Zoppini G, et al. (1993) Mononuclear leukocytes from obese patients with type II diabetes have reduced activity of hexokinase, 6-phosphofructokinase and glucose-6-phosphate dehydrogenase. *Horm Metab Res* 25: 160-164.

[110] Gaskin RS, Estwick D, Peddi R (2001) G6PD deficiency: its role in the high prevalence of hypertension and diabetes mellitus. *Ethn Dis* 11: 749-754.

[111] Cappai G, Songini M, Doria A, Cavallerano JD, Lorenzi M (2011) Increased prevalence of proliferative retinopathy in patients with type 1 diabetes who are deficient in glucose-6-phosphate dehydrogenase. *Diabetologia* 54: 1539-1542.

[112] Tian WN, Braunstein LD, Apse K, Pang J, Rose M, et al. (1999) Importance of glucose-6-phosphate dehydrogenase activity in cell death. *Am J Physiol* 276: C1121-1131.

[113] Leopold JA, Zhang YY, Scribner AW, Stanton RC, Loscalzo J (2003) Glucose-6-phosphate dehydrogenase overexpression decreases endothelial cell oxidant stress and increases bioavailable nitric oxide. *Arterioscler Thromb Vasc Bio*l 23: 411-417.

[114] Pan S, World CJ, Kovacs CJ, Berk BC (2009) Glucose 6-phosphate dehydrogenase is regulated through c-Src-mediated tyrosine phosphorylation in endothelial cells. *Arterioscler Thromb Vasc Biol* 29: 895-901.

[115] Fildes RA, Parr CW (1963) Human red-cell phosphogluconate dehydrogenases. *Nature* 200: 890-891.

[116] Carter ND, Fildes RA, Fitch LI, Parr CW (1968) Genetically determined electrophoretic variations of human phosphogluconate dehydrogenase. *Acta Genet Stat Med* 18: 109-122.

[117] Tuchinda S, Rucknagel DL, Na-Nakorn S, Wasi P (1968) The Thai variant and the distribution of alleles of 6-phosphogluconate dehydrogenase and the distribution of glucose 6-phosphate dehydrogenase deficiency in Thailand. *Biochem Genet* 2: 253-264.

[118] Brewer GJ, Dern RJ (1964) A New Inherited enzymatic deficiency of human erythrocytes: 6-phosphogluconate dehydrogenase deficiency. *Am J Hum Genet* 16: 472-476.

[119] Parr CW, Fitch LI (1964) Hereditary partial deficiency of human-erythrocyte phosphogluconate dehydrogenase. *Biochem J* 93: 28C-30C.

[120] Ajmar F, Lamedica G, Garre C, Ravazzolo R, Sessarego M, et al. (1979) Partial deficiency of red cell 6-phosphogluconate dehydrogenase: a family study. *Hum Genet* 52: 347-351.

[121] Caprari P, Caforio MP, Cianciulli P, Maffi D, Pasquino MT, et al. (2001) 6-Phosphogluconate dehydrogenase deficiency in an Italian family. *Ann Hematol* 80: 41-44.

[122] Lausecker C, Hartleyb H, Fischer D, Roujon J (1965) Massive quinine poisoning. Treatment by exsanguinotransfusion. Electrocardiographic study. *Arch Fr Pediatr* 22: 1008-1019.

[123] Scialom C, Najean Y, Bernard J (1966) Congenital nonspherocytic hemolytic anemia with incomplete deficiency of 6-phosphogluconate dehydrogenase. *Nouv Rev Fr Hematol* 6: 452-457.

[124] Chayen J, Bitensky L, Aves EK, Jones GR, Silcox AA, et al. (1962) Histochemical demonstration of 6-phosphogluconate dehydrogenase in proliferating and malignant cells. *Nature* 195: 714-715.

[125] Bonham DG, Gibbs DF (1962) A new enzyme test for gynaecological cancer. 6-Phosphogluconate dehydrogenase activity in vaginal fluid. *Br Med J* 2: 823-824.

[126] Muir GG, Canti G, Williams D (1964) Use of 6-phosphogluconate dehydrogenase as a screen test for cervical carcinoma in normal women. *Br Med J* 2: 1563-1565.

[127] Cohen S, Way S (1966) Histochemical demonstration of pentose shunt activity in smears from the uterine cervix. *Br Med J* 1: 88-89.

[128] Goldberg DM, Watts C, Hart DM (1970) Evaluation of several enzyme tests in vaginal fluid as aids to the diagnosis of invasive and preinvasive cervical cancer. *Am J Obstet Gynecol* 107: 465-471.

[129] Rees KR, Slater TF, Gibbs DF, Stagg BH (1970) A modified assay for 6-phosphogluconate dehydrogenase in samples of vaginal fluid from women with and without gynecologic cancer. *Am J Obstet Gynecol* 107: 857-864.

[130] Jonas SK, Benedetto C, Flatman A, Hammond RH, Micheletti L, et al. (1992) Increased activity of 6-phosphogluconate dehydrogenase and glucose-6-phosphate dehydrogenase in purified cell suspensions and single cells from the uterine cervix in cervical intraepithelial neoplasia. *Br J Cancer* 66: 185-191.

[131] Denton JE, Lui MS, Aoki T, Sebolt J, Takeda E, et al. (1982) Enzymology of pyrimidine and carbohydrate metabolism in human colon carcinomas. *Cancer Res* 42: 1176-1183.

[132] Jonges GN, Vogels IM, Bosch KS, Dingemans KP, Van Noorden CJ (1993) Experimentally induced colon cancer metastases in rat liver increase the proliferation rate and capacity for purine catabolism in liver cells. *Histochemistry* 100: 41-51.

[133] Sukhatme VP, Chan B (2012) Glycolytic cancer cells lacking 6-phosphogluconate dehydrogenase metabolize glucose to induce senescence. *FEBS Lett* 586: 2389-2395.

[134] Chan B, VanderLaan PA, Sukhatme VP (2013) 6-Phosphogluconate dehydrogenase regulates tumor cell migration *in vitro* by regulating receptor tyrosine kinase c-Met. *Biochem Biophys Res Commun* 439: 247-251.

[135] Shan C, Elf S, Ji Q, Kang HB, Zhou L, et al. (2014) Lysine acetylation activates 6-phosphogluconate dehydrogenase to promote tumor growth. *Mol Cell* 55: 552-565.

[136] Lin R, Elf S, Shan C, Kang HB, Ji Q, et al. (2015) 6-Phosphogluconate dehydrogenase links oxidative PPP, lipogenesis and tumour growth by inhibiting LKB1-AMPK signalling. *Nat Cell Biol* 17: 1484-1496.

[137] Gray LR, Tompkins SC, Taylor EB (2014) Regulation of pyruvate metabolism and human disease. *Cell Mol Life Sci* 71: 2577-2604.

[138] Patel MS, Korotchkina LG (2006) Regulation of the pyruvate dehydrogenase complex. *Biochem Soc Trans* 34: 217-222.

[139] Cai Z, Zhao JS, Li JJ, Peng DN, Wang XY, et al. (2010) A combined proteomics and metabolomics profiling of gastric cardia cancer reveals characteristic dysregulations in glucose metabolism. *Mol Cell Proteomics* 9: 2617-2628.

[140] Sun XR, Sun Z, Zhu Z, Guan HX, Li CY, et al. (2015) Expression of pyruvate dehydrogenase is an independent prognostic marker in gastric cancer. *World J Gastroenterol* 21: 5336-5344.

[141] Zhong Y, Huang R, Li X, Xu R, Zhou F, et al. (2015) Decreased expression of PDHE1 α predicts worse clinical outcome in esophageal squamous cell carcinoma. *Anticancer Res* 35: 5533-5538.

[142] Goh WQ, Ow GS, Kuznetsov VA, Chong S, Lim YP (2015) DLAT subunit of the pyruvate dehydrogenase complex is upregulated in gastric cancer-implications in cancer therapy. *Am J Transl Res* 7: 1140-1151.

[143] Xu XD, Shao SX, Jiang HP, Cao YW, Wang YH, et al. (2015) Warburg effect or reverse Warburg effect? A review of cancer metabolism. *Oncol Res Treat* 38: 117-122.

[144] Glushakova LG, Lisankie MJ, Eruslanov EB, Ojano-Dirain C, Zolotukhin I, et al. (2009) AAV3-mediated transfer and expression of the pyruvate dehydrogenase E1α subunit gene causes metabolic remodeling and apoptosis of human liver cancer cells. *Mol Genet Metab* 98: 289-299.

[145] McFate T, Mohyeldin A, Lu H, Thakar J, Henriques J, et al. (2008) Pyruvate dehydrogenase complex activity controls metabolic and malignant phenotype in cancer cells. *J Biol Chem* 283: 22700-22708.

[146] Sidhu S, Gangasani A, Korotchkina LG, Suzuki G, Fallavollita JA, et al. (2008) Tissue-specific pyruvate dehydrogenase complex deficiency causes cardiac hypertrophy and sudden death of weaned male mice. *Am J Physiol Heart Circ Physiol* 295: H946-H952.

[147] Zhao G, Jeoung NH, Burgess SC, Rosaaen-Stowe KA, Inagaki T, et al. (2008) Overexpression of pyruvate dehydrogenase kinase 4 in heart perturbs metabolism and exacerbates calcineurin-induced cardiomyopathy. *Am J Physiol Heart Circ Physiol* 294: H936-943.

[148] Lewandowski ED, White LT (1995) Pyruvate dehydrogenase influences postischemic heart function. *Circulation* 91: 2071-2079.

[149] Bersin RM, Stacpoole PW (1997) Dichloroacetate as metabolic therapy for myocardial ischemia and failure. *Am Heart J* 134: 841-855.

[150] White LT, O'Donnell JM, Griffin J, Lewandowski ED (1999) Cytosolic redox state mediates postischemic response to pyruvate dehydrogenase stimulation. *Am J Physiol* 277: H626-634.

[151] Wada N, Matsuishi T, Nonaka M, Naito E, Yoshino M (2004) Pyruvate dehydrogenase E1α subunit deficiency in a female patient: evidence of antenatal origin of brain damage and possible etiology of infantile spasms. *Brain Dev* 26: 57-60.

[152] Tamaru S, Kikuchi A, Takagi K, Okuno J, Ishikawa K, et al. (2012) A case of pyruvate dehydrogenase E1α subunit deficiency with antenatal brain dysgenesis demonstrated by prenatal sonography and magnetic resonance imaging. *J Clin Ultrasound* 40: 234-238.

[153] Small JE, Gonzalez GE, Nagao KE, Walton DS, Caruso PA (2009) Optic neuropathy in a patient with pyruvate dehydrogenase deficiency. *Pediatr Radiol* 39: 1114-1117.

[154] Barnerias C, Saudubray JM, Touati G, De Lonlay P, Dulac O, et al. (2010) Pyruvate dehydrogenase complex deficiency: four neurological phenotypes with differing pathogenesis. *Dev Med Child Neurol* 52: e1-9.

[155] DeBrosse SD, Okajima K, Zhang S, Nakouzi G, Schmotzer CL, et al. (2012) Spectrum of neurological and survival outcomes in pyruvate dehydrogenase complex (PDC) deficiency: lack of correlation with genotype. *Mol Genet Metab* 107: 394-402.

[156] Sharma P, Benford B, Li ZZ, Ling GS (2009) Role of pyruvate dehydrogenase complex in traumatic brain injury and measurement of pyruvate dehydrogenase enzyme by dipstick test. *J Emerg Trauma Shock* 2: 67-72.

[157] Pliss L, Hausknecht KA, Stachowiak MK, Dlugos CA, Richards JB, et al. (2013) Cerebral developmental abnormalities in a mouse with systemic pyruvate dehydrogenase deficiency. *PLoS One* 8: e67473.

[158] Srinivasan M, Choi CS, Ghoshal P, Pliss L, Pandya JD, et al. (2010) ss-Cell-specific pyruvate dehydrogenase deficiency impairs glucose-stimulated insulin secretion. *Am J Physiol Endocrinol Metab* 299: E910-917.

[159] Nellemann B, Vendelbo MH, Nielsen TS, Bak AM, Hogild M, et al. (2014) Growth hormone-induced insulin resistance in human subjects involves reduced pyruvate dehydrogenase activity. *Acta Physiol (Oxf)* 210: 392-402.

[160] Jeoung NH, Harris RA (2008) Pyruvate dehydrogenase kinase-4 deficiency lowers blood glucose and improves glucose tolerance in diet-induced obese mice. *Am J Physiol Endocrinol Metab* 295: E46-54.

[161] Andersen LW, Berg KM, Montissol S, Sulmonte C, Balkema J, et al. (2016) Pyruvate dehydrogenase activity is decreased in emergency department patients with diabetic ketoacidosis. *Acad Emerg Med* 23: 685-689.

[162] Markert CL, Shaklee JB, Whitt GS (1975) Evolution of a gene. Multiple genes for LDH isozymes provide a model of the evolution of gene structure, function and regulation. *Science* 189: 102-114.

[163] Rong Y, Wu W, Ni X, Kuang T, Jin D, et al. (2013) Lactate dehydrogenase A is overexpressed in pancreatic cancer and promotes the growth of pancreatic cancer cells. *Tumour Biol* 34: 1523-1530.

[164] Zhao D, Zou SW, Liu Y, Zhou X, Mo Y, et al. (2013) Lysine-5 acetylation negatively regulates lactate dehydrogenase A and is decreased in pancreatic cancer. *Cancer Cell* 23: 464-476.

[165] Shi M, Cui J, Du J, Wei D, Jia Z, et al. (2014) A novel KLF4/LDHA signaling pathway regulates aerobic glycolysis in and progression of pancreatic cancer. *Clin Cancer Res* 20: 4370-4380.

[166] Cui J, Shi M, Xie D, Wei D, Jia Z, et al. (2014) FOXM1 promotes the Warburg effect and pancreatic cancer progression via transactivation of LDHA expression. *Clin Cancer Res* 20: 2595-2606.

[167] He C, Jiang H, Geng S, Sheng H, Shen X, et al. (2014) Expression and prognostic value of c-Myc and Fas (CD95/APO1) in patients with pancreatic cancer. *Int J Clin Exp Pathol* 7: 742-750.

[168] He TL, Zhang YJ, Jiang H, Li XH, Zhu H, et al. (2015) The c-Myc-LDHA axis positively regulates aerobic glycolysis and promotes tumor progression in pancreatic cancer. *Med Oncol* 32: 187.

[169] Cui J, Quan M, Jiang W, Hu H, Jiao F, et al. (2015) Suppressed expression of LDHB promotes pancreatic cancer progression via inducing glycolytic phenotype. *Med Oncol* 32: 143.

[170] Hussien R, Brooks GA (2011) Mitochondrial and plasma membrane lactate transporter and lactate dehydrogenase isoform expression in breast cancer cell lines. *Physiol Genomics* 43: 255-264.

[171] Agrawal A, Gandhe MB, Gupta D, Reddy MV (2016) Preliminary study on serum lactate dehydrogenase (LDH)-prognostic biomarker in carcinoma breast. *J Clin Diagn Res* 10: BC06-08.

[172] Liu X, Meng QH, Ye Y, Hildebrandt MA, Gu J, et al. (2015) Prognostic significance of pretreatment serum levels of albumin, LDH and total bilirubin in patients with non-metastatic breast cancer. *Carcinogenesis* 36: 243-248.

[173] Brown JE, Cook RJ, Lipton A, Coleman RE (2012) Serum lactate dehydrogenase is prognostic for survival in patients with bone metastases from breast cancer: a retrospective analysis in bisphosphonate-treated patients. *Clin Cancer Res* 18: 6348-6355.

[174] Wang ZY, Loo TY, Shen JG, Wang N, Wang DM, et al. (2012) LDH-A silencing suppresses breast cancer tumorigenicity through induction of oxidative stress mediated mitochondrial pathway apoptosis. *Breast Cancer Res Treat* 131: 791-800.

[175] Rizwan A, Serganova I, Khanin R, Karabeber H, Ni X, et al. (2013) Relationships between LDH-A, lactate, and metastases in 4T1 breast tumors. *Clin Cancer Res* 19: 5158-5169.

[176] Xiao X, Huang X, Ye F, Chen B, Song C, et al. (2016) The miR-34a-LDHA axis regulates glucose metabolism and tumor growth in breast cancer. *Sci Rep* 6: 21735.

[177] Brown NJ, Higham SE, Perunovic B, Arafa M, Balasubramanian S, et al. (2013) Lactate dehydrogenase-B is silenced by promoter methylation in a high frequency of human breast cancers. *PLoS One* 8: e57697.

[178] Arora R, Schmitt D, Karanam B, Tan M, Yates C, et al. (2015) Inhibition of the Warburg effect with a natural compound reveals a novel measurement for determining the metastatic potential of breast cancers. *Oncotarget* 6: 662-678.

[179] McCleland ML, Adler AS, Shang Y, Hunsaker T, Truong T, et al. (2012) An integrated genomic screen identifies LDHB as an essential gene for triple-negative breast cancer. *Cancer Res* 72: 5812-5823.

[180] Dennison JB, Molina JR, Mitra S, Gonzalez-Angulo AM, Balko JM, et al. (2013) Lactate dehydrogenase B: a metabolic marker of response to

neoadjuvant chemotherapy in breast cancer. *Clin Cancer Res* 19: 3703-3713.

[181] Kim JH, Kim EL, Lee YK, Park CB, Kim BW, et al. (2011) Decreased lactate dehydrogenase B expression enhances claudin 1-mediated hepatoma cell invasiveness via mitochondrial defects. *Exp Cell Res* 317: 1108-1118.

[182] Mirebeau-Prunier D, Le Pennec S, Jacques C, Fontaine JF, Gueguen N, et al. (2013) Estrogen-related receptor α modulates lactate dehydrogenase activity in thyroid tumors. *PLoS One* 8: e58683.

[183] Koukourakis MI, Giatromanolaki A, Simopoulos C, Polychronidis A, Sivridis E (2005) Lactate dehydrogenase 5 (LDH5) relates to up-regulated hypoxia inducible factor pathway and metastasis in colorectal cancer. *Clin Exp Metastasis* 22: 25-30.

[184] Kachel P, Trojanowicz B, Sekulla C, Prenzel H, Dralle H, et al. (2015) Phosphorylation of pyruvate kinase M2 and lactate dehydrogenase A by fibroblast growth factor receptor 1 in benign and malignant thyroid tissue. *BMC Cancer* 15: 140.

[185] Kolev Y, Uetake H, Takagi Y, Sugihara K (2008) Lactate dehydrogenase-5 (LDH-5) expression in human gastric cancer: association with hypoxia-inducible factor (HIF-1α) pathway, angiogenic factors production and poor prognosis. *Ann Surg Oncol* 15: 2336-2344.

[186] Zhao Z, Han F, Yang S, Hua L, Wu J, et al. (2014) The clinicopathologic importance of serum lactic dehydrogenase in patients with gastric cancer. *Dis Markers* 2014: 140913

[187] Zhao Z, Han F, Yang S, Wu J, Zhan W (2015) Oxamate-mediated inhibition of lactate dehydrogenase induces protective autophagy in gastric cancer cells: involvement of the Akt-mTOR signaling pathway. *Cancer Lett* 358: 17-26.

[188] Zha X, Wang F, Wang Y, He S, Jing Y, et al. (2011) Lactate dehydrogenase B is critical for hyperactive mTOR-mediated tumorigenesis. *Cancer Res* 71: 13-18.

[189] Daifuku M, Nishi K, Okamoto T, Sugahara T (2014) Activation of J774.1 murine macrophages by lactate dehydrogenase. *Cytotechnology* 66: 937-943.

[190] Ahn H, Lee K, Kim JM, Kwon SH, Lee SH, et al. (2016) Accelerated lactate dehydrogenase activity potentiates osteoclastogenesis via NFATc1 signaling. *PLoS One* 11: e0153886.

[191] Chung KM, Chang ST, Huang WT, Lu CL, Wu HC, et al. (2013) Bcl-6 expression and lactate dehydrogenase level predict prognosis of primary

gastric diffuse large B-cell lymphoma. *J Formos Med Assoc* 112: 382-389.

[192] Egberts F, Kotthoff EM, Gerdes S, Egberts JH, Weichenthal M, et al. (2012) Comparative study of YKL-40, S-100B and LDH as monitoring tools for Stage IV melanoma. *Eur J Cancer* 48: 695-702.

[193] Cetin B, Afsar B, Deger SM, Gonul, II, Gumusay O, et al. (2014) Association between hemoglobin, calcium, and lactate dehydrogenase variability and mortality among metastatic renal cell carcinoma. *Int Urol Nephrol* 46: 1081-1087.

[194] Sun X, Sun Z, Zhu Z, Li C, Zhang J, et al. (2015) Expression of SIP1 is strongly correlated with LDHA and shows a significantly poor outcome in gastric cancer. *Tumour Biol* 36: 7521-7530.

[195] Koukourakis MI, Kakouratos C, Kalamida D, Bampali Z, Mavropoulou S, et al. (2016) Hypoxia-inducible proteins HIF-1α and lactate dehydrogenase LDH5, key markers of anaerobic metabolism, relate with stem cell markers and poor post-radiotherapy outcome in bladder cancer. *Int J Radiat Biol* 92: 353-363.

[196] Kanemasa Y, Shimoyama T, Sasaki Y, Sawada T, Omuro Y, et al. (2016) A convenient prognostic score consisting of the Glasgow prognostic score and serum lactate dehydrogenase predicts clinical outcome in patients with diffuse large B-cell lymphoma. *Leuk Lymphoma* 57: 2460-2463.

[197] Kim HS, Lee HE, Yang HK, Kim WH (2014) High lactate dehydrogenase 5 expression correlates with high tumoral and stromal vascular endothelial growth factor expression in gastric cancer. *Pathobiology* 81: 78-85.

[198] Zhuang L, Scolyer RA, Murali R, McCarthy SW, Zhang XD, et al. (2010) Lactate dehydrogenase 5 expression in melanoma increases with disease progression and is associated with expression of Bcl-x/l and Mcl-1, but not Bcl-2 proteins. *Mod Pathol* 23: 45-53.

[199] Passardi A, Scarpi E, Tamberi S, Cavanna L, Tassinari D, et al. (2015) Impact of pre-treatment lactate dehydrogenase levels on prognosis and bevacizumab efficacy in patients with metastatic colorectal cancer. *PLoS One* 10: e0134732.

[200] Mohammad GH, Olde Damink SW, Malago M, Dhar DK, Pereira SP (2016) Pyruvate kinase M2 and lactate dehydrogenase A are overexpressed in pancreatic cancer and correlate with poor outcome. *PLoS One* 11: e0151635.

[201] Faloppi L, Bianconi M, Giampieri R, Sobrero A, Labianca R, et al. (2015) The value of lactate dehydrogenase serum levels as a prognostic and predictive factor for advanced pancreatic cancer patients receiving sorafenib. *Oncotarget* 6: 35087-35094.

[202] D'Cruz AM, Pathiyil V (2015) Histopathological differentiation of oral squamous cell carcinoma and salivary lactate dehydrogenase: A biochemical study. *South Asian J Cancer* 4: 58-60.

[203] Saluja TS, Spadigam A, Dhupar A, Syed S (2016) Equating salivary lactate dehydrogenase (LDH) with LDH-5 expression in patients with oral squamous cell carcinoma: An insight into metabolic reprogramming of cancer cell as a predictor of aggressive phenotype. *Tumour Biol* 37: 5609-5620.

[204] Grimm M, Alexander D, Munz A, Hoffmann J, Reinert S (2013) Increased LDH5 expression is associated with lymph node metastasis and outcome in oral squamous cell carcinoma. *Clin Exp Metastasis* 30: 529-540.

[205] Cunningham D, Humblet Y, Siena S, Khayat D, Bleiberg H, et al. (2004) Cetuximab monotherapy and cetuximab plus irinotecan in irinotecan-refractory metastatic colorectal cancer. *N Engl J Med* 351: 337-345.

[206] Bonner JA, Harari PM, Giralt J, Azarnia N, Shin DM, et al. (2006) Radiotherapy plus cetuximab for squamous-cell carcinoma of the head and neck. *N Engl J Med* 354: 567-578.

[207] Wu X, Fan Z, Masui H, Rosen N, Mendelsohn J (1995) Apoptosis induced by an anti-epidermal growth factor receptor monoclonal antibody in a human colorectal carcinoma cell line and its delay by insulin. *J Clin Invest* 95: 1897-1905.

[208] Lu H, Li X, Luo Z, Liu J, Fan Z (2013) Cetuximab reverses the Warburg effect by inhibiting HIF-1-regulated LDH-A. *Mol Cancer Ther* 12: 2187-2199.

[209] Le A, Cooper CR, Gouw AM, Dinavahi R, Maitra A, et al. (2010) Inhibition of lactate dehydrogenase A induces oxidative stress and inhibits tumor progression. *Proc Natl Acad Sci U S A* 107: 2037-2042.

[210] Liu X, Yang Z, Chen Z, Chen R, Zhao D, et al. (2015) Effects of the suppression of lactate dehydrogenase A on the growth and invasion of human gastric cancer cells. *Oncol Rep* 33: 157-162.

[211] Manerba M, Vettraino M, Fiume L, Di Stefano G, Sartini A, et al. (2012) Galloflavin (CAS 568-80-9): a novel inhibitor of lactate dehydrogenase. *ChemMedChem* 7: 311-317.

[212] Farabegoli F, Vettraino M, Manerba M, Fiume L, Roberti M, et al. (2012) Galloflavin, a new lactate dehydrogenase inhibitor, induces the death of human breast cancer cells with different glycolytic attitude by affecting distinct signaling pathways. *Eur J Pharm Sci* 47: 729-738.

[213] Vettraino M, Manerba M, Govoni M, Di Stefano G (2013) Galloflavin suppresses lactate dehydrogenase activity and causes MYC downregulation in Burkitt lymphoma cells through NAD/NADH-dependent inhibition of sirtuin-1. *Anticancer Drugs* 24: 862-870.

[214] Fiume L, Vettraino M, Carnicelli D, Arfilli V, Di Stefano G, et al. (2013) Galloflavin prevents the binding of lactate dehydrogenase A to single stranded DNA and inhibits RNA synthesis in cultured cells. *Biochem Biophys Res Commun* 430: 466-469.

[215] Han X, Sheng X, Jones HM, Jackson AL, Kilgore J, et al. (2015) Evaluation of the anti-tumor effects of lactate dehydrogenase inhibitor galloflavin in endometrial cancer cells. *J Hematol Oncol* 8: 2.

[216] Lu NN, Weng ZY, Chen QY, Boison D, Xiao XX, et al. (2016) Evaluation on the inhibition of pyrrol-2-yl ethanone derivatives to lactate dehydrogenase and anticancer activities. *Spectrochim Acta A Mol Biomol Spectrosc* 165: 21-25.

[217] Mokwatsi GG, Schutte AE, Kruger R (2016) A biomarker of tissue damage, lactate dehydrogenase, is associated with fibulin-1 and oxidative stress in blacks: the SAfrEIC study. *Biomarkers* 21: 48-55.

[218] Parra-Bonilla G, Alvarez DF, Al-Mehdi AB, Alexeyev M, Stevens T (2010) Critical role for lactate dehydrogenase A in aerobic glycolysis that sustains pulmonary microvascular endothelial cell proliferation. *Am J Physiol Lung Cell Mol Physiol* 299: L513-522.

[219] Buckner SL, Loenneke JP, Loprinzi PD (2016) Cross-sectional association between normal-range lactate dehydrogenase, physical activity and cardiovascular disease risk score. *Sports Med* 46: 467-472.

[220] Bo G, Hui L, Feng-Li L, Cui L (2012) Relationships between edema degree and clinical and biochemical parameters in posterior reversible encephalopathy syndrome: a preliminary study. *Acta Neurol Belg* 112: 281-285.

[221] Schwartz RB, Feske SK, Polak JF, DeGirolami U, Iaia A, et al. (2000) Preeclampsia-eclampsia: clinical and neuroradiographic correlates and insights into the pathogenesis of hypertensive encephalopathy. *Radiology* 217: 371-376.

[222] Demirtas O, Gelal F, Vidinli BD, Demirtas LO, Uluc E, et al. (2005) Cranial MR imaging with clinical correlation in preeclampsia and eclampsia. *Diagn Interv Radiol* 11: 189-194.

[223] Kato GJ, McGowan V, Machado RF, Little JA, Taylor Jt, et al. (2006) Lactate dehydrogenase as a biomarker of hemolysis-associated nitric oxide resistance, priapism, leg ulceration, pulmonary hypertension, and death in patients with sickle cell disease. *Blood* 107: 2279-2285.

[224] Mikobi TM, Lukusa Tshilobo P, Aloni MN, Mvumbi Lelo G, Akilimali PZ, et al. (2015) Correlation between the lactate dehydrogenase levels with laboratory variables in the clinical severity of sickle cell anemia in Congolese patients. *PLoS One* 10: e0123568.

[225] Bosron WF, Li TK (1986) Genetic polymorphism of human liver alcohol and aldehyde dehydrogenases, and their relationship to alcohol metabolism and alcoholism. *Hepatology* 6: 502-510.

[226] Bosron WF, Li TK (1987) Catalytic properties of human liver alcohol dehydrogenase isoenzymes. *Enzyme* 37: 19-28.

[227] Ehrig T, Bosron WF, Li TK (1990) Alcohol and aldehyde dehydrogenase. *Alcohol Alcohol* 25: 105-116.

[228] Pares X, Vallee BL (1981) New human liver alcohol dehydrogenase forms with unique kinetic characteristics. *Biochem Biophys Res Commun* 98: 122-130.

[229] Yin SJ, Han CL, Lee AI, Wu CW (1999) Human alcohol dehydrogenase family. Functional classification, ethanol/retinol metabolism, and medical implications. *Adv Exp Med Biol* 463: 265-271.

[230] Moreno A, Pares X (1991) Purification and characterization of a new alcohol dehydrogenase from human stomach. *J Biol Chem* 266: 1128-1133.

[231] Crabb DW, Matsumoto M, Chang D, You M (2004) Overview of the role of alcohol dehydrogenase and aldehyde dehydrogenase and their variants in the genesis of alcohol-related pathology. *Proc Nutr Soc* 63: 49-63.

[232] Edenberg HJ (2007) The genetics of alcohol metabolism: role of alcohol dehydrogenase and aldehyde dehydrogenase variants. *Alcohol Res Health* 30: 5-13.

[233] Osier MV, Pakstis AJ, Goldman D, Edenberg HJ, Kidd JR, et al. (2002) A proline-threonine substitution in codon 351 of ADH1C is common in Native Americans. *Alcohol Clin Exp Res* 26: 1759-1763.

[234] Osier MV, Pakstis AJ, Soodyall H, Comas D, Goldman D, et al. (2002) A global perspective on genetic variation at the ADH genes reveals

unusual patterns of linkage disequilibrium and diversity. *Am J Hum Genet* 71: 84-99.
[235] Stickel F, Osterreicher CH (2006) The role of genetic polymorphisms in alcoholic liver disease. *Alcohol Alcohol* 41: 209-224.
[236] Deltour L, Foglio MH, Duester G (1999) Metabolic deficiencies in alcohol dehydrogenase Adh1, Adh3, and Adh4 null mutant mice. Overlapping roles of Adh1 and Adh4 in ethanol clearance and metabolism of retinol to retinoic acid. *J Biol Chem* 274: 16796-16801.
[237] Yin SJ, Chou FJ, Chao SF, Tsai SF, Liao CS, et al. (1993) Alcohol and aldehyde dehydrogenases in human esophagus: comparison with the stomach enzyme activities. *Alcohol Clin Exp Res* 17: 376-381.
[238] Dong YJ, Peng TK, Yin SJ (1996) Expression and activities of class IV alcohol dehydrogenase and class III aldehyde dehydrogenase in human mouth. *Alcohol* 13: 257-262.
[239] Peng GS, Yin SJ (2009) Effect of the allelic variants of aldehyde dehydrogenase *ALDH2*2* and alcohol dehydrogenase *ADH1B*2* on blood acetaldehyde concentrations. *Hum Genomics* 3: 121-127.
[240] Whitfield JB (1997) Meta-analysis of the effects of alcohol dehydrogenase genotype on alcohol dependence and alcoholic liver disease. *Alcohol Alcohol* 32: 613-619.
[241] Zintzaras E, Stefanidis I, Santos M, Vidal F (2006) Do alcohol-metabolizing enzyme gene polymorphisms increase the risk of alcoholism and alcoholic liver disease? *Hepatology* 43: 352-361.
[242] Borras E, Coutelle C, Rosell A, Fernandez-Muixi F, Broch M, et al. (2000) Genetic polymorphism of alcohol dehydrogenase in europeans: the *ADH2*2* allele decreases the risk for alcoholism and is associated with *ADH3*1*. *Hepatology* 31: 984-989.
[243] Li D, Zhao H, Gelernter J (2011) Strong association of the alcohol dehydrogenase 1B gene (*ADH1B*) with alcohol dependence and alcohol-induced medical diseases. *Biol Psychiatry* 70: 504-512.
[244] Ferrari P, McKay JD, Jenab M, Brennan P, Canzian F, et al. (2012) Alcohol dehydrogenase and aldehyde dehydrogenase gene polymorphisms, alcohol intake and the risk of colorectal cancer in the European Prospective Investigation into Cancer and Nutrition study. *Eur J Clin Nutr* 66: 1303-1308.
[245] Neumark YD, Friedlander Y, Durst R, Leitersdorf E, Jaffe D, et al. (2004) Alcohol dehydrogenase polymorphisms influence alcohol-elimination rates in a male Jewish population. *Alcohol Clin Exp Res* 28: 10-14.

[246] Yokoyama A, Yokoyama T, Mizukami T, Matsui T, Kimura M, et al. (2016) Alcohol dehydrogenase-1B (rs1229984) and aldehyde dehydrogenase-2 (rs671) genotypes and alcoholic ketosis are associated with the serum uric acid level in Japanese alcoholic men. *Alcohol Alcohol* 51: 268-274.

[247] Edenberg HJ, Xuei X, Chen HJ, Tian H, Wetherill LF, et al. (2006) Association of alcohol dehydrogenase genes with alcohol dependence: a comprehensive analysis. *Hum Mol Genet* 15: 1539-1549.

[248] Kuo PH, Kalsi G, Prescott CA, Hodgkinson CA, Goldman D, et al. (2008) Association of *ADH* and *ALDH* genes with alcohol dependence in the Irish Affected Sib Pair Study of alcohol dependence (IASPSAD) sample. *Alcohol Clin Exp Res* 32: 785-795.

[249] Roy N, Dasgupta D, Mukhopadhyay I, Chatterjee A, Das K, et al. (2016) Genetic association and gene-gene interaction reveal genetic variations in *ADH1B*, *GSTM1* and *MnSOD* independently confer risk to alcoholic liver diseases in India. *PLoS One* 11: e0149843.

[250] McCarver DG (2001) *ADH2* and *CYP2E1* genetic polymorphisms: risk factors for alcohol-related birth defects. *Drug Metab Dispos* 29: 562-565.

[251] Khan AJ, Husain Q, Choudhuri G, Parmar D (2010) Association of polymorphism in alcohol dehydrogenase and interaction with other genetic risk factors with alcoholic liver cirrhosis. *Drug Alcohol Depend* 109: 190-197.

[252] Toth R, Fiatal S, Petrovski B, McKee M, Adany R (2011) Combined effect of *ADH1B* RS1229984, RS2066702 and *ADH1C* RS1693482/ RS698 alleles on alcoholism and chronic liver diseases. *Dis Markers* 31: 267-277.

[253] Osier M, Pakstis AJ, Kidd JR, Lee JF, Yin SJ, et al. (1999) Linkage disequilibrium at the *ADH2* and *ADH3* loci and risk of alcoholism. *Am J Hum Genet* 64: 1147-1157.

[254] Osier MV, Lu RB, Pakstis AJ, Kidd JR, Huang SY, et al. (2004) Possible epistatic role of ADH7 in the protection against alcoholism. *Am J Med Genet B Neuropsychiatr Genet* 126B: 19-22.

[255] Yamauchi M, Takamatsu M, Maezawa Y, Takagi M, Araki T, et al. (2001) Polymorphism of tumor necrosis factor-beta and alcohol dehydrogenase genes and alcoholic brain atrophy in Japanese patients. *Alcohol Clin Exp Res* 25: 7S-10S.

[256] Buervenich S, Sydow O, Carmine A, Zhang Z, Anvret M, et al. (2000) Alcohol dehydrogenase alleles in Parkinson's disease. *Mov Disord* 15: 813-818.

[257] Buervenich S, Carmine A, Galter D, Shahabi HN, Johnels B, et al. (2005) A rare truncating mutation in *ADH1C* (G78Stop) shows significant association with Parkinson disease in a large international sample. *Arch Neurol* 62: 74-78.

[258] Belin AC, Westerlund M, Anvret A, Lindqvist E, Pernold K, et al. (2011) Modeling Parkinson's disease genetics: altered function of the dopamine system in Adh4 knockout mice. *Behav Brain Res* 217: 439-445.

[259] Anvret A, Ran C, Westerlund M, Gellhaar S, Lindqvist E, et al. (2012) Adh1 and Adh1/4 knockout mice as possible rodent models for presymptomatic Parkinson's disease. *Behav Brain Res* 227: 252-257.

[260] Hines LM, Rimm EB (2001) Moderate alcohol consumption and coronary heart disease: a review. *Postgrad Med J* 77: 747-752.

[261] Djousse L, Levy D, Herbert AG, Wilson PW, D'Agostino RB, et al. (2005) Influence of alcohol dehydrogenase 1C polymorphism on the alcohol-cardiovascular disease association (from the Framingham Offspring Study). *Am J Cardiol* 96: 227-232.

[262] Younis J, Cooper JA, Miller GJ, Humphries SE, Talmud PJ (2005) Genetic variation in alcohol dehydrogenase 1C and the beneficial effect of alcohol intake on coronary heart disease risk in the Second Northwick Park Heart Study. *Atherosclerosis* 180: 225-232.

[263] Chen YJ, Chen C, Wu DC, Lee CH, Wu CI, et al. (2006) Interactive effects of lifetime alcohol consumption and alcohol and aldehyde dehydrogenase polymorphisms on esophageal cancer risks. *Int J Cancer* 119: 2827-2831.

[264] Cui R, Kamatani Y, Takahashi A, Usami M, Hosono N, et al. (2009) Functional variants in *ADH1B* and *ALDH2* coupled with alcohol and smoking synergistically enhance esophageal cancer risk. *Gastroenterology* 137: 1768-1775.

[265] Akbari MR, Malekzadeh R, Shakeri R, Nasrollahzadeh D, Foumani M, et al. (2009) Candidate gene association study of esophageal squamous cell carcinoma in a high-risk region in Iran. *Cancer Res* 69: 7994-8000.

[266] Wu C, Kraft P, Zhai K, Chang J, Wang Z, et al. (2012) Genome-wide association analyses of esophageal squamous cell carcinoma in Chinese identify multiple susceptibility loci and gene-environment interactions. *Nat Genet* 44: 1090-1097.

[267] Wang J, Wei J, Xu X, Pan W, Ge Y, et al. (2014) Replication study of ESCC susceptibility genetic polymorphisms locating in the *ADH1B-ADH1C-ADH7* cluster identified by GWAS. *PLoS One* 9: e94096.

[268] Wu M, Chang SC, Kampman E, Yang J, Wang XS, et al. (2013) Single nucleotide polymorphisms of *ADH1B*, *ADH1C* and *ALDH2* genes and esophageal cancer: a population-based case-control study in China. *Int J Cancer* 132: 1868-1877.

[269] Terry MB, Gammon MD, Zhang FF, Vaughan TL, Chow WH, et al. (2007) Alcohol dehydrogenase 3 and risk of esophageal and gastric adenocarcinomas. *Cancer Causes Control* 18: 1039-1046.

[270] Israel Y, Rivera-Meza M, Quintanilla ME, Sapag A, Tampier L (2011) Acetaldehyde burst protection of *ADH1B*2* against alcoholism: an additional hormesis protection against esophageal cancers following alcohol consumption? *Alcohol Clin Exp Res* 35: 806-810.

[271] Feinendegen LE (2005) Evidence for beneficial low level radiation effects and radiation hormesis. *Br J Radiol* 78: 3-7.

[272] Parsons PA (1990) Radiation hormesis: an evolutionary expectation and the evidence. *Int J Rad Appl Instrum A* 41: 857-860.

[273] Willett WC, Stampfer MJ, Colditz GA, Rosner BA, Hennekens CH, et al. (1987) Moderate alcohol consumption and the risk of breast cancer. *N Engl J Med* 316: 1174-1180.

[274] Smith-Warner SA, Spiegelman D, Yaun SS, van den Brandt PA, Folsom AR, et al. (1998) Alcohol and breast cancer in women: a pooled analysis of cohort studies. *JAMA* 279: 535-540.

[275] Feron VJ, Til HP, de Vrijer F, Woutersen RA, Cassee FR, et al. (1991) Aldehydes: occurrence, carcinogenic potential, mechanism of action and risk assessment. *Mutat Res* 259: 363-385.

[276] Ristow H, Seyfarth A, Lochmann ER (1995) Chromosomal damages by ethanol and acetaldehyde in *Saccharomyces cerevisiae* as studied by pulsed field gel electrophoresis. *Mutat Res* 326: 165-170.

[277] Freudenheim JL, Ambrosone CB, Moysich KB, Vena JE, Graham S, et al. (1999) Alcohol dehydrogenase 3 genotype modification of the association of alcohol consumption with breast cancer risk. *Cancer Causes Control* 10: 369-377.

[278] Terry MB, Gammon MD, Zhang FF, Knight JA, Wang Q, et al. (2006) *ADH3* genotype, alcohol intake and breast cancer risk. *Carcinogenesis* 27: 840-847.

[279] Bernstein L, Ross RK (1993) Endogenous hormones and breast cancer risk. *Epidemiol Rev* 15: 48-65.

[280] Ginsburg E, Vonderhaar BK (1995) Prolactin synthesis and secretion by human breast cancer cells. *Cancer Res* 55: 2591-2595.

[281] Purohit V (1998) Moderate alcohol consumption and estrogen levels in postmenopausal women: a review. *Alcohol Clin Exp Res* 22: 994-997.

[282] Trichopoulos D, Brown J, MacMahon B (1987) Urine estrogens and breast cancer risk factors among post-menopausal women. *Int J Cancer* 40: 721-725.

[283] Hines LM, Hankinson SE, Smith-Warner SA, Spiegelman D, Kelsey KT, et al. (2000) A prospective study of the effect of alcohol consumption and *ADH3* genotype on plasma steroid hormone levels and breast cancer risk. *Cancer Epidemiol Biomarkers Prev* 9: 1099-1105.

[284] Peters ES, McClean MD, Liu M, Eisen EA, Mueller N, et al. (2005) The *ADH1C* polymorphism modifies the risk of squamous cell carcinoma of the head and neck associated with alcohol and tobacco use. *Cancer Epidemiol Biomarkers Prev* 14: 476-482.

[285] Wang D, Ritchie JM, Smith EM, Zhang Z, Turek LP, et al. (2005) Alcohol dehydrogenase 3 and risk of squamous cell carcinomas of the head and neck. *Cancer Epidemiol Biomarkers Prev* 14: 626-632.

[286] Olshan AF, Weissler MC, Watson MA, Bell DA (2001) Risk of head and neck cancer and the alcohol dehydrogenase 3 genotype. *Carcinogenesis* 22: 57-61.

[287] Brennan P, Lewis S, Hashibe M, Bell DA, Boffetta P, et al. (2004) Pooled analysis of alcohol dehydrogenase genotypes and head and neck cancer: a HuGE review. *Am J Epidemiol* 159: 1-16.

[288] Chang JS, Straif K, Guha N (2012) The role of alcohol dehydrogenase genes in head and neck cancers: a systematic review and meta-analysis of *ADH1B* and *ADH1C*. *Mutagenesis* 27: 275-286.

[289] Zhang Y, Gu N, Miao L, Yuan H, Wang R, et al. (2015) Alcohol dehydrogenase-1B Arg47His polymorphism is associated with head and neck cancer risk in Asian: a meta-analysis. *Tumour Biol* 36: 1023-1027.

[290] Ji YB, Lee SH, Kim KR, Park CW, Song CM, et al. (2015) Association between *ADH1B* and *ADH1C* polymorphisms and the risk of head and neck squamous cell carcinoma. *Tumour Biol* 36: 4387-4396.

[291] Minegishi Y, Tsukino H, Muto M, Goto K, Gemma A, et al. (2007) Susceptibility to lung cancer and genetic polymorphisms in the alcohol metabolite-related enzymes alcohol dehydrogenase 3, aldehyde dehydrogenase 2, and cytochrome P450 2E1 in the Japanese population. *Cancer* 110: 353-362.

[292] Mutka SC, Green LH, Verderber EL, Richards JP, Looker DL, et al. (2012) *ADH IB* expression, but not *ADH III*, is decreased in human lung cancer. *PLoS One* 7: e52995.

[293] Freudenheim JL, Ram M, Nie J, Muti P, Trevisan M, et al. (2003) Lung cancer in humans is not associated with lifetime total alcohol consumption or with genetic variation in alcohol dehydrogenase 3 (*ADH3*). *J Nutr* 133: 3619-3624.

[294] Obe G, Jonas R, Schmidt S (1986) Metabolism of ethanol *in vitro* produces a compound which induces sister-chromatid exchanges in human peripheral lymphocytes *in vitro*: acetaldehyde not ethanol is mutagenic. *Mutat Res* 174: 47-51.

[295] Jelski W, Zalewski B, Chrostek L, Szmitkowski M (2007) Alcohol dehydrogenase (ADH) isoenzymes and aldehyde dehydrogenase (ALDH) activity in the sera of patients with colorectal cancer. *Clin Exp Med* 7: 154-157.

[296] Tiemersma EW, Wark PA, Ocke MC, Bunschoten A, Otten MH, et al. (2003) Alcohol consumption, alcohol dehydrogenase 3 polymorphism, and colorectal adenomas. *Cancer Epidemiol Biomarkers Prev* 12: 419-425.

[297] Homann N, Konig IR, Marks M, Benesova M, Stickel F, et al. (2009) Alcohol and colorectal cancer: the role of alcohol dehydrogenase 1C polymorphism. *Alcohol Clin Exp Res* 33: 551-556.

[298] Homann N, Stickel F, Konig IR, Jacobs A, Junghanns K, et al. (2006) Alcohol dehydrogenase 1C*1 allele is a genetic marker for alcohol-associated cancer in heavy drinkers. *Int J Cancer* 118: 1998-2002.

[299] Xue Y, Wang M, Zhong D, Tong N, Chu H, et al. (2012) ADH1C Ile350Val polymorphism and cancer risk: evidence from 35 case-control studies. *PLoS One* 7: e37227.

[300] Jelski W, Kutylowska E, Szmitkowski M (2011) The role of ethanol in pathogenesis of pancreatitis. *Pol Merkur Lekarski* 30: 66-68.

[301] Sturgis EM, Dahlstrom KR, Guan Y, Eicher SA, Strom SS, et al. (2001) Alcohol dehydrogenase 3 genotype is not associated with risk of squamous cell carcinoma of the oral cavity and pharynx. *Cancer Epidemiol Biomarkers Prev* 10: 273-275.

[302] Singh D, Negi TS, Upadhyay G, Choudhuri G (2015) Polymorphism of alcohol metabolizing gene *ADH3* predisposes to development of alcoholic pancreatitis in North Indian population. *Front Mol Biosci* 2: 67.

[303] Frenzer A, Butler WJ, Norton ID, Wilson JS, Apte MV, et al. (2002) Polymorphism in alcohol-metabolizing enzymes, glutathione S-transferases and apolipoprotein E and susceptibility to alcohol-induced cirrhosis and chronic pancreatitis. *J Gastroenterol Hepatol* 17: 177-182.

[304] Verlaan M, Te Morsche RH, Roelofs HM, Laheij RJ, Jansen JB, et al. (2004) Genetic polymorphisms in alcohol-metabolizing enzymes and chronic pancreatitis. *Alcohol Alcohol* 39: 20-24.

[305] Fang F, Pan J, Su GH, Xu LX, Li G, et al. (2015) Association between alcohol dehydrogenase 1C gene *1/*2 polymorphism and pancreatitis risk: a meta-analysis. *Genet Mol Res* 14: 15267-15275.

[306] Goto M, Kitamura H, Alam MM, Ota N, Haseba T, et al. (2015) Alcohol dehydrogenase 3 contributes to the protection of liver from nonalcoholic steatohepatitis. *Genes Cells* 20: 464-480.

[307] Yan H, Parsons DW, Jin G, McLendon R, Rasheed BA, et al. (2009) IDH1 and IDH2 mutations in gliomas. *N Engl J Med* 360: 765-773.

[308] Marcucci G, Maharry K, Wu YZ, Radmacher MD, Mrozek K, et al. (2010) *IDH1* and *IDH2* gene mutations identify novel molecular subsets within de novo cytogenetically normal acute myeloid leukemia: a Cancer and Leukemia Group B study. *J Clin Oncol* 28: 2348-2355.

[309] Murugan AK, Bojdani E, Xing M (2010) Identification and functional characterization of isocitrate dehydrogenase 1 (IDH1) mutations in thyroid cancer. *Biochem Biophys Res Commun* 393: 555-559.

[310] Ward PS, Cross JR, Lu C, Weigert O, Abel-Wahab O, et al. (2012) Identification of additional IDH mutations associated with oncometabolite R(-)-2-hydroxyglutarate production. *Oncogene* 31: 2491-2498.

[311] Dang L, White DW, Gross S, Bennett BD, Bittinger MA, et al. (2009) Cancer-associated IDH1 mutations produce 2-hydroxyglutarate. *Nature* 462: 739-744.

[312] Sjoblom T, Jones S, Wood LD, Parsons DW, Lin J, et al. (2006) The consensus coding sequences of human breast and colorectal cancers. *Science* 314: 268-274.

[313] Parsons DW, Jones S, Zhang X, Lin JC, Leary RJ, et al. (2008) An integrated genomic analysis of human glioblastoma multiforme. *Science* 321: 1807-1812.

[314] Balss J, Meyer J, Mueller W, Korshunov A, Hartmann C, et al. (2008) Analysis of the IDH1 codon 132 mutation in brain tumors. *Acta Neuropathol* 116: 597-602.

[315] Mardis ER, Ding L, Dooling DJ, Larson DE, McLellan MD, et al. (2009) Recurring mutations found by sequencing an acute myeloid leukemia genome. *N Engl J Med* 361: 1058-1066.

[316] Molenaar RJ, Thota S, Nagata Y, Patel B, Clemente M, et al. (2015) Clinical and biological implications of ancestral and non-ancestral IDH1 and IDH2 mutations in myeloid neoplasms. *Leukemia* 29: 2134-2142.

[317] Fathi AT, Sadrzadeh H, Comander AH, Higgins MJ, Bardia A, et al. (2014) Isocitrate dehydrogenase 1 (IDH1) mutation in breast adenocarcinoma is associated with elevated levels of serum and urine 2-hydroxyglutarate. *Oncologist* 19: 602-607.

[318] Wang C, McKeithan TW, Gong Q, Zhang W, Bouska A, et al. (2015) IDH2R172 mutations define a unique subgroup of patients with angioimmunoblastic T-cell lymphoma. *Blood* 126: 1741-1752.

[319] Amary MF, Bacsi K, Maggiani F, Damato S, Halai D, et al. (2011) IDH1 and IDH2 mutations are frequent events in central chondrosarcoma and central and periosteal chondromas but not in other mesenchymal tumours. *J Pathol* 224: 334-343.

[320] Borger DR, Zhu AX (2012) IDH mutations: new genetic signatures in cholangiocarcinoma and therapeutic implications. *Expert Rev Anticancer Ther* 12: 543-546.

[321] Kipp BR, Voss JS, Kerr SE, Barr Fritcher EG, Graham RP, et al. (2012) Isocitrate dehydrogenase 1 and 2 mutations in cholangiocarcinoma. *Hum Pathol* 43: 1552-1558.

[322] Snuderl M, Triscott J, Northcott PA, Shih HA, Kong E, et al. (2015) Deep sequencing identifies IDH1 R132S mutation in adult medulloblastoma. *J Clin Oncol* 33: e27-31.

[323] Hartman DJ, Binion D, Regueiro M, Schraut W, Bahary N, et al. (2014) Isocitrate dehydrogenase-1 is mutated in inflammatory bowel disease-associated intestinal adenocarcinoma with low-grade tubuloglandular histology but not in sporadic intestinal adenocarcinoma. *Am J Surg Pathol* 38: 1147-1156.

[324] Whitehall VL, Dumenil TD, McKeone DM, Bond CE, Bettington ML, et al. (2014) Isocitrate dehydrogenase 1 R132C mutation occurs exclusively in microsatellite stable colorectal cancers with the CpG island methylator phenotype. *Epigenetics* 9: 1454-1460.

[325] Liu X, Kato Y, Kaneko MK, Sugawara M, Ogasawara S, et al. (2013) Isocitrate dehydrogenase 2 mutation is a frequent event in osteosarcoma detected by a multi-specific monoclonal antibody MsMab-1. *Cancer Med* 2: 803-814.

[326] Chotirat S, Thongnoppakhun W, Promsuwicha O, Boonthimat C, Auewarakul CU (2012) Molecular alterations of isocitrate dehydrogenase 1 and 2 (IDH1 and IDH2) metabolic genes and additional genetic mutations in newly diagnosed acute myeloid leukemia patients. *J Hematol Oncol* 5: 5.

[327] Cairns RA, Iqbal J, Lemonnier F, Kucuk C, de Leval L, et al. (2012) IDH2 mutations are frequent in angioimmunoblastic T-cell lymphoma. *Blood* 119: 1901-1903.

[328] Xu X, Zhao J, Xu Z, Peng B, Huang Q, et al. (2004) Structures of human cytosolic NADP-dependent isocitrate dehydrogenase reveal a novel self-regulatory mechanism of activity. *J Biol Chem* 279: 33946-33957.

[329] Kolker S, Pawlak V, Ahlemeyer B, Okun JG, Horster F, et al. (2002) NMDA receptor activation and respiratory chain complex V inhibition contribute to neurodegeneration in d-2-hydroxyglutaric aciduria. *Eur J Neurosci* 16: 21-28.

[330] Latini A, Scussiato K, Rosa RB, Llesuy S, Bello-Klein A, et al. (2003) D-2-hydroxyglutaric acid induces oxidative stress in cerebral cortex of young rats. *Eur J Neurosci* 17: 2017-2022.

[331] Aghili M, Zahedi F, Rafiee E (2009) Hydroxyglutaric aciduria and malignant brain tumor: a case report and literature review. *J Neurooncol* 91: 233-236.

[332] Mussini E, Hutton JJ, Jr., Udenfriend S (1967) Collagen proline hydroxylase in wound healing, granuloma formation, scurvy, and growth. *Science* 157: 927-929.

[333] Hausinger RP (2004) FeII/α-ketoglutarate-dependent hydroxylases and related enzymes. *Crit Rev Biochem Mol Biol* 39: 21-68.

[334] Loenarz C, Schofield CJ (2008) Expanding chemical biology of 2-oxoglutarate oxygenases. *Nat Chem Biol* 4: 152-156.

[335] Xu W, Yang H, Liu Y, Yang Y, Wang P, et al. (2011) Oncometabolite 2-hydroxyglutarate is a competitive inhibitor of α-ketoglutarate-dependent dioxygenases. *Cancer Cell* 19: 17-30.

[336] Chowdhury R, Yeoh KK, Tian YM, Hillringhaus L, Bagg EA, et al. (2011) The oncometabolite 2-hydroxyglutarate inhibits histone lysine demethylases. *EMBO Rep* 12: 463-469.

[337] Luksch H, Uckermann O, Stepulak A, Hendruschk S, Marzahn J, et al. (2011) Silencing of selected glutamate receptor subunits modulates cancer growth. *Anticancer Res* 31: 3181-3192.

[338] Figueroa ME, Abdel-Wahab O, Lu C, Ward PS, Patel J, et al. (2010) Leukemic IDH1 and IDH2 mutations result in a hypermethylation

phenotype, disrupt TET2 function, and impair hematopoietic differentiation. *Cancer Cell* 18: 553-567.

[339] Ito S, D'Alessio AC, Taranova OV, Hong K, Sowers LC, et al. (2010) Role of Tet proteins in 5mC to 5hmC conversion, ES-cell self-renewal and inner cell mass specification. *Nature* 466: 1129-1133.

[340] Guo JU, Su Y, Zhong C, Ming GL, Song H (2011) Hydroxylation of 5-methylcytosine by TET1 promotes active DNA demethylation in the adult brain. *Cell* 145: 423-434.

[341] Guilhamon P, Eskandarpour M, Halai D, Wilson GA, Feber A, et al. (2013) Meta-analysis of IDH-mutant cancers identifies EBF1 as an interaction partner for TET2. *Nat Commun* 4: 2166.

[342] Kooistra SM, Helin K (2012) Molecular mechanisms and potential functions of histone demethylases. *Nat Rev Mol Cell Biol* 13: 297-311.

[343] Lu C, Ward PS, Kapoor GS, Rohle D, Turcan S, et al. (2012) IDH mutation impairs histone demethylation and results in a block to cell differentiation. *Nature* 483: 474-478.

[344] Turcan S, Rohle D, Goenka A, Walsh LA, Fang F, et al. (2012) IDH1 mutation is sufficient to establish the glioma hypermethylator phenotype. *Nature* 483: 479-483.

[345] Duncan CG, Barwick BG, Jin G, Rago C, Kapoor-Vazirani P, et al. (2012) A heterozygous IDH1R132H/WT mutation induces genome-wide alterations in DNA methylation. *Genome Res* 22: 2339-2355.

[346] Noushmehr H, Weisenberger DJ, Diefes K, Phillips HS, Pujara K, et al. (2010) Identification of a CpG island methylator phenotype that defines a distinct subgroup of glioma. *Cancer Cell* 17: 510-522.

[347] van den Bent MJ, Dubbink HJ, Marie Y, Brandes AA, Taphoorn MJ, et al. (2010) IDH1 and IDH2 mutations are prognostic but not predictive for outcome in anaplastic oligodendroglial tumors: a report of the European Organization for Research and Treatment of Cancer Brain Tumor Group. *Clin Cancer Res* 16: 1597-1604.

[348] Ivanova AV, Ivanov SV, Danilkovitch-Miagkova A, Lerman MI (2001) Regulation of STRA13 by the von Hippel Lindau tumor suppressor protein, hypoxia, and the UBC9/ubiquitin proteasome degradation pathway. *J Biol Chem* 276: 15306-15315.

[349] Zhao S, Lin Y, Xu W, Jiang W, Zha Z, et al. (2009) Glioma-derived mutations in IDH1 dominantly inhibit IDH1 catalytic activity and induce HIF-1α. *Science* 324: 261-265.

[350] Zhu J, Cui G, Chen M, Xu Q, Wang X, et al. (2013) Expression of R132H mutational IDH1 in human U87 glioblastoma cells affects the

SREBP1a pathway and induces cellular proliferation. *J Mol Neurosci* 50: 165-171.
[351] Sasaki M, Knobbe CB, Itsumi M, Elia AJ, Harris IS, et al. (2012) D-2-hydroxyglutarate produced by mutant IDH1 perturbs collagen maturation and basement membrane function. *Genes Dev* 26: 2038-2049.
[352] Reitman ZJ, Jin G, Karoly ED, Spasojevic I, Yang J, et al. (2011) Profiling the effects of isocitrate dehydrogenase 1 and 2 mutations on the cellular metabolome. *Proc Natl Acad Sci U S A* 108: 3270-3275.
[353] Izquierdo-Garcia JL, Cai LM, Chaumeil MM, Eriksson P, Robinson AE, et al. (2014) Glioma cells with the IDH1 mutation modulate metabolic fractional flux through pyruvate carboxylase. *PLoS One* 9: e108289.
[354] Izquierdo-Garcia JL, Viswanath P, Eriksson P, Cai L, Radoul M, et al. (2015) IDH1 mutation induces reprogramming of pyruvate metabolism. *Cancer Res* 75: 2999-3009.
[355] Grassian AR, Parker SJ, Davidson SM, Divakaruni AS, Green CR, et al. (2014) IDH1 mutations alter citric acid cycle metabolism and increase dependence on oxidative mitochondrial metabolism. *Cancer Res* 74: 3317-3331.
[356] Rohle D, Popovici-Muller J, Palaskas N, Turcan S, Grommes C, et al. (2013) An inhibitor of mutant IDH1 delays growth and promotes differentiation of glioma cells. *Science* 340: 626-630.
[357] Wang F, Travins J, DeLaBarre B, Penard-Lacronique V, Schalm S, et al. (2013) Targeted inhibition of mutant IDH2 in leukemia cells induces cellular differentiation. *Science* 340: 622-626.
[358] DiNardo CD, Ravandi F, Agresta S, Konopleva M, Takahashi K, et al. (2015) Characteristics, clinical outcome, and prognostic significance of IDH mutations in AML. *Am J Hematol* 90: 732-736.
[359] Davis M, Pragani R, Popovici-Muller J, Gross S, Thorne N, et al. (2010) ML309: A potent inhibitor of R132H mutant IDH1 capable of reducing 2-hydroxyglutarate production in U87 MG glioblastoma cells. *Probe Reports from the NIH Molecular Libraries Program*. Bethesda (MD).
[360] Kim HJ, Choi BY, Keum YS (2015) Identification of a new selective chemical inhibitor of mutant isocitrate dehydrogenase-1. *J Cancer Prev* 20: 78-83.
[361] Li W, Zhao Y (2013) Withaferin A suppresses tumor promoter 12-O-tetradecanoylphorbol 13-acetate-induced decreases in isocitrate dehydrogenase 1 activity and mitochondrial function in skin epidermal JB6 cells. *Cancer Sci* 104: 143-148.

[362] Chan SM, Thomas D, Corces-Zimmerman MR, Xavy S, Rastogi S, et al. (2015) Isocitrate dehydrogenase 1 and 2 mutations induce Bcl-2 dependence in acute myeloid leukemia. *Nat Med* 21: 178-184.

[363] Schumacher T, Bunse L, Wick W, Platten M (2014) Mutant IDH1: An immunotherapeutic target in tumors. *Oncoimmunology* 3: e974392.

[364] Gross S, Cairns RA, Minden MD, Driggers EM, Bittinger MA, et al. (2010) Cancer-associated metabolite 2-hydroxyglutarate accumulates in acute myelogenous leukemia with isocitrate dehydrogenase 1 and 2 mutations. *J Exp Med* 207: 339-344.

[365] Ward PS, Patel J, Wise DR, Abdel-Wahab O, Bennett BD, et al. (2010) The common feature of leukemia-associated IDH1 and IDH2 mutations is a neomorphic enzyme activity converting α-ketoglutarate to 2-hydroxyglutarate. *Cancer Cell* 17: 225-234.

[366] Aydin K, Ozmen M, Tatli B, Sencer S (2003) Single-voxel MR spectroscopy and diffusion-weighted MRI in two patients with l-2-hydroxyglutaric aciduria. *Pediatr Radiol* 33: 872-876.

[367] Tanaka K, Sasayama T, Mizukawa K, Takata K, Sulaiman NS, et al. (2015) Combined IDH1 mutation and MGMT methylation status on long-term survival of patients with cerebral low-grade glioma. *Clin Neurol Neurosurg* 138: 37-44.

[368] Dai Y, Ning X, Han G, Li W (2016) Assessment of the association between isocitrate dehydrogenase 1 mutation and mortality risk of glioblastoma patients. *Mol Neurobiol* 53: 1501-1508.

[369] Miyata S, Urabe M, Gomi A, Nagai M, Yamaguchi T, et al. (2013) An R132H mutation in isocitrate dehydrogenase 1 enhances p21 expression and inhibits phosphorylation of retinoblastoma protein in glioma cells. *Neurol Med Chir (Tokyo)* 53: 645-654.

[370] Zhu J, Zuo J, Xu Q, Wang X, Wang Z, et al. (2011) Isocitrate dehydrogenase mutations may be a protective mechanism in glioma patients. *Med Hypotheses* 76: 602-603.

[371] Shen Y, Zhu YM, Fan X, Shi JY, Wang QR, et al. (2011) Gene mutation patterns and their prognostic impact in a cohort of 1185 patients with acute myeloid leukemia. *Blood* 118: 5593-5603.

[372] Ma QL, Wang JH, Wang YG, Hu C, Mu QT, et al. (2015) High IDH1 expression is associated with a poor prognosis in cytogenetically normal acute myeloid leukemia. *Int J Cancer* 137: 1058-1065.

[373] Zhang H, Guo X, Dai J, Wu Y, Ge N, et al. (2014) Genetic variations in IDH gene as prognosis predictors in TACE-treated hepatocellular carcinoma patients. *Med Oncol* 31: 278.

[374] Tan F, Jiang Y, Sun N, Chen Z, Lv Y, et al. (2012) Identification of isocitrate dehydrogenase 1 as a potential diagnostic and prognostic biomarker for non-small cell lung cancer by proteomic analysis. *Mol Cell Proteomics* 11: M111 008821.

[375] Guay C, Joly E, Pepin E, Barbeau A, Hentsch L, et al. (2013) A role for cytosolic isocitrate dehydrogenase as a negative regulator of glucose signaling for insulin secretion in pancreatic ss-cells. *PLoS One* 8: e77097.

[376] Yoshimi N, Futamura T, Bergen SE, Iwayama Y, Ishima T, et al. (2016) Cerebrospinal fluid metabolomics identifies a key role of isocitrate dehydrogenase in bipolar disorder: evidence in support of mitochondrial dysfunction hypothesis. *Mol Psychiatry*. 1-7.

[377] Zhang D, Wang Y, Shi Z, Liu J, Sun P, et al. (2015) Metabolic reprogramming of cancer-associated fibroblasts by IDH3α downregulation. Cell Rep 10: 1335-1348.

[378] Gibson GE, Park LC, Sheu KF, Blass JP, Calingasan NY (2000) The α-ketoglutarate dehydrogenase complex in neurodegeneration. *Neurochem Int* 36: 97-112.

[379] Sheu KF, Blass JP (1999) The α-ketoglutarate dehydrogenase complex. *Ann N Y Acad Sci* 893: 61-78.

[380] Hensley CT, Wasti AT, DeBerardinis RJ (2013) Glutamine and cancer: cell biology, physiology, and clinical opportunities. *J Clin Invest* 123: 3678-3684.

[381] Starkov AA (2013) An update on the role of mitochondrial α-ketoglutarate dehydrogenase in oxidative stress. *Mol Cell Neurosci* 55: 13-16.

[382] Tretter L, Adam-Vizi V (2000) Inhibition of Krebs cycle enzymes by hydrogen peroxide: A key role of α-ketoglutarate dehydrogenase in limiting NADH production under oxidative stress. *J Neurosci* 20: 8972-8979.

[383] Nulton-Persson AC, Szweda LI (2001) Modulation of mitochondrial function by hydrogen peroxide. *J Biol Chem* 276: 23357-23361.

[384] Humphries KM, Szweda LI (1998) Selective inactivation of α-ketoglutarate dehydrogenase and pyruvate dehydrogenase: reaction of lipoic acid with 4-hydroxy-2-nonenal. *Biochemistry* 37: 15835-15841.

[385] Applegate MA, Humphries KM, Szweda LI (2008) Reversible inhibition of α-ketoglutarate dehydrogenase by hydrogen peroxide: glutathionylation and protection of lipoic acid. *Biochemistry* 47: 473-478.

[386] Tretter L, Adam-Vizi V (2004) Generation of reactive oxygen species in the reaction catalyzed by α-ketoglutarate dehydrogenase. *J Neurosci* 24: 7771-7778.

[387] Starkov AA, Fiskum G, Chinopoulos C, Lorenzo BJ, Browne SE, et al. (2004) Mitochondrial α-ketoglutarate dehydrogenase complex generates reactive oxygen species. *J Neurosci* 24: 7779-7788.

[388] Zundorf G, Kahlert S, Bunik VI, Reiser G (2009) α-Ketoglutarate dehydrogenase contributes to production of reactive oxygen species in glutamate-stimulated hippocampal neurons in situ. *Neuroscience* 158: 610-616.

[389] Santos SS, Gibson GE, Cooper AJ, Denton TT, Thompson CM, et al. (2006) Inhibitors of the α-ketoglutarate dehydrogenase complex alter [1-^{13}C]glucose and [U-^{13}C]glutamate metabolism in cerebellar granule neurons. *J Neurosci Res* 83: 450-458.

[390] Nilsen LH, Shi Q, Gibson GE, Sonnewald U (2011) Brain [U-^{13}C]glucose metabolism in mice with decreased α-ketoglutarate dehydrogenase complex activity. *J Neurosci Res* 89: 1997-2007.

[391] Shi Q, Risa O, Sonnewald U, Gibson GE (2009) Mild reduction in the activity of the α-ketoglutarate dehydrogenase complex elevates GABA shunt and glycolysis. *J Neurochem* 109 Suppl 1: 214-221.

[392] Kuroda Y, Kline JJ, Sweetman L, Nyhan WL, Groshong TD (1979) Abnormal pyruvate and α-ketoglutarate dehydrogenase complexes in a patient with lactic acidemia. *Pediatr Res* 13: 928-931.

[393] Dunckelmann RJ, Ebinger F, Schulze A, Wanders RJ, Rating D, et al. (2000) 2-ketoglutarate dehydrogenase deficiency with intermittent 2-ketoglutaric aciduria. *Neuropediatrics* 31: 35-38.

[394] Huang HM, Zhang H, Xu H, Gibson GE (2003) Inhibition of the α-ketoglutarate dehydrogenase complex alters mitochondrial function and cellular calcium regulation. *Biochim Biophys Acta* 1637: 119-126.

[395] Calingasan NY, Ho DJ, Wille EJ, Campagna MV, Ruan J, et al. (2008) Influence of mitochondrial enzyme deficiency on adult neurogenesis in mouse models of neurodegenerative diseases. *Neuroscience* 153: 986-996.

[396] Weinert BT, Scholz C, Wagner SA, Iesmantavicius V, Su D, et al. (2013) Lysine succinylation is a frequently occurring modification in prokaryotes and eukaryotes and extensively overlaps with acetylation. *Cell Rep* 4: 842-851.

[397] Gibson GE, Xu H, Chen HL, Chen W, Denton TT, et al. (2015) α-Ketoglutarate dehydrogenase complex-dependent succinylation of proteins in neurons and neuronal cell lines. *J Neurochem* 134: 86-96.
[398] Mastrogiacoma F, Lindsay JG, Bettendorff L, Rice J, Kish SJ (1996) Brain protein and α-ketoglutarate dehydrogenase complex activity in Alzheimer's disease. *Ann Neurol* 39: 592-598.
[399] Gibson GE, Sheu KF, Blass JP (1998) Abnormalities of mitochondrial enzymes in Alzheimer disease. *J Neural Transm (Vienna)* 105: 855-870.
[400] Sheu KF, Brown AM, Haroutunian V, Kristal BS, Thaler H, et al. (1999) Modulation by DLST of the genetic risk of Alzheimer's disease in a very elderly population. *Ann Neurol* 45: 48-53.
[401] Nakano K, Ohta S, Nishimaki K, Miki T, Matuda S (1997) Alzheimer's disease and DLST genotype. *Lancet* 350: 1367-1368.
[402] Matsushita S, Arai H, Yuzuriha T, Kato M, Matsui T, et al. (2001) No association between DLST gene and Alzheimer's disease or Wernicke-Korsakoff syndrome. *Neurobiol Aging* 22: 569-574.
[403] Brown AM, Gordon D, Lee H, Caudy M, Hardy J, et al. (2004) Association of the dihydrolipoamide dehydrogenase gene with Alzheimer's disease in an Ashkenazi Jewish population. *Am J Med Genet B Neuropsychiatr Genet* 131B: 60-66.
[404] Mizuno Y, Ikebe S, Hattori N, Nakagawa-Hattori Y, Mochizuki H, et al. (1995) Role of mitochondria in the etiology and pathogenesis of Parkinson's disease. *Biochim Biophys Acta* 1271: 265-274.
[405] Lương Kvqc, Nguyên LTH (2012) Thiamine and Parkinson's disease. *J Neurol Sci* 316: 1-8.
[406] Joffe GT, Parks JK, Parker WD, Jr. (1998) Secondary inhibition of 2-ketoglutarate dehydrogenase complex by MPTP. *Neuroreport* 9: 2781-2783.
[407] Mizuno Y, Matuda S, Yoshino H, Mori H, Hattori N, et al. (1994) An immunohistochemical study on α-ketoglutarate dehydrogenase complex in Parkinson's disease. *Ann Neurol* 35: 204-210.
[408] Kobayashi T, Matsumine H, Matuda S, Mizuno Y (1998) Association between the gene encoding the E2 subunit of the α-ketoglutarate dehydrogenase complex and Parkinson's disease. *Ann Neurol* 43: 120-123.
[409] Fedorova MS, Kudryavtseva AV, Lakunina VA, Snezhkina AV, Volchenko NN, et al. (2015) Downregulation of OGDHL expression is associated with promoter hypermethylation in colorectal cancer. *Mol Biol (Mosk)* 49: 678-688.

[410] Sen T, Sen N, Noordhuis MG, Ravi R, Wu TC, et al. (2012) OGDHL is a modifier of AKT-dependent signaling and NF-κB function. *PLoS One* 7: e48770.

[411] Bunik VI, Mkrtchyan G, Grabarska A, Oppermann H, Daloso D, et al. (2016) Inhibition of mitochondrial 2-oxoglutarate dehydrogenase impairs viability of cancer cells in a cell-specific metabolism-dependent manner. *Oncotarget* 7: 26400-26421.

[412] Cecchini G (2003) Function and structure of complex II of the respiratory chain. *Annu Rev Biochem* 72: 77-109.

[413] Rutter J, Winge DR, Schiffman JD (2010) Succinate dehydrogenase - assembly, regulation and role in human disease. *Mitochondrion* 10: 393-401.

[414] Bardella C, Pollard PJ, Tomlinson I (2011) SDH mutations in cancer. *Biochim Biophys Acta* 1807: 1432-1443.

[415] Neumann HP, Pawlu C, Peczkowska M, Bausch B, McWhinney SR, et al. (2004) Distinct clinical features of paraganglioma syndromes associated with *SDHB* and *SDHD* gene mutations. *JAMA* 292: 943-951.

[416] Havekes B, Corssmit EP, Jansen JC, van der Mey AG, Vriends AH, et al. (2007) Malignant paragangliomas associated with mutations in the succinate dehydrogenase D gene. *J Clin Endocrinol Metab* 92: 1245-1248.

[417] Bennedbaek M, Rossing M, Rasmussen AK, Gerdes AM, Skytte AB, et al. (2016) Identification of eight novel *SDHB*, *SDHC*, *SDHD* germline variants in Danish pheochromocytoma/paraganglioma patients. *Hered Cancer Clin Pract* 14: 13.

[418] Hao HX, Khalimonchuk O, Schraders M, Dephoure N, Bayley JP, et al. (2009) *SDH5*, a gene required for flavination of succinate dehydrogenase, is mutated in paraganglioma. *Science* 325: 1139-1142.

[419] Bayley JP, Kunst HP, Cascon A, Sampietro ML, Gaal J, et al. (2010) *SDHAF2* mutations in familial and sporadic paraganglioma and phaeochromocytoma. *Lancet Oncol* 11: 366-372.

[420] Kunst HP, Rutten MH, de Monnink JP, Hoefsloot LH, Timmers HJ, et al. (2011) *SDHAF2* (PGL2-SDH5) and hereditary head and neck paraganglioma. *Clin Cancer Res* 17: 247-254.

[421] Niemann S, Muller U, Engelhardt D, Lohse P (2003) Autosomal dominant malignant and catecholamine-producing paraganglioma caused by a splice donor site mutation in *SDHC*. *Hum Genet* 113: 92-94.

[422] Millar AC, Mete O, Cusimano RJ, Fremes SE, Keshavjee S, et al. (2014) Functional cardiac paraganglioma associated with a rare *SDHC* mutation. *Endocr Pathol* 25: 315-320.
[423] Bayley JP, van Minderhout I, Weiss MM, Jansen JC, Oomen PH, et al. (2006) Mutation analysis of *SDHB* and *SDHC*: novel germline mutations in sporadic head and neck paraganglioma and familial paraganglioma and/or pheochromocytoma. *BMC Med Genet* 7: 1.
[424] Ricketts CJ, Forman JR, Rattenberry E, Bradshaw N, Lalloo F, et al. (2010) Tumor risks and genotype-phenotype-proteotype analysis in 358 patients with germline mutations in *SDHB* and *SDHD*. *Hum Mutat* 31: 41-51.
[425] Rich T, Jackson M, Roman-Gonzalez A, Shah K, Cote GJ, et al. (2015) Metastatic sympathetic paraganglioma in a patient with loss of the *SDHC* gene. *Fam Cancer* 14: 615-619.
[426] Else T, Marvin ML, Everett JN, Gruber SB, Arts HA, et al. (2014) The clinical phenotype of SDHC-associated hereditary paraganglioma syndrome (PGL3). *J Clin Endocrinol Metab* 99: E1482-1486.
[427] Astuti D, Latif F, Dallol A, Dahia PL, Douglas F, et al. (2001) Gene mutations in the succinate dehydrogenase subunit *SDHB* cause susceptibility to familial pheochromocytoma and to familial paraganglioma. *Am J Hum Genet* 69: 49-54.
[428] Peterson LA, Litzendorf M, Ringel MD, Vaccaro PS (2014) *SDHB* gene mutation in a carotid body paraganglioma: case report and review of the paraganglioma syndromes. *Ann Vasc Surg* 28: 1321 e1329-1312.
[429] Burnichon N, Briere JJ, Libe R, Vescovo L, Riviere J, et al. (2010) *SDHA* is a tumor suppressor gene causing paraganglioma. *Hum Mol Genet* 19: 3011-3020.
[430] Mason EF, Sadow PM, Wagner AJ, Remillard SP, Flood TA, et al. (2013) Identification of succinate dehydrogenase-deficient bladder paragangliomas. *Am J Surg Pathol* 37: 1612-1618.
[431] Dwight T, Mann K, Benn DE, Robinson BG, McKelvie P, et al. (2013) Familial *SDHA* mutation associated with pituitary adenoma and pheochromocytoma/paraganglioma. *J Clin Endocrinol Metab* 98: E1103-1108.
[432] Miettinen M, Wang ZF, Sarlomo-Rikala M, Osuch C, Rutkowski P, et al. (2011) Succinate dehydrogenase-deficient GISTs: a clinicopathologic, immunohistochemical, and molecular genetic study of 66 gastric GISTs with predilection to young age. *Am J Surg Pathol* 35: 1712-1721.

[433] Janeway KA, Kim SY, Lodish M, Nose V, Rustin P, et al. (2011) Defects in succinate dehydrogenase in gastrointestinal stromal tumors lacking KIT and PDGFRA mutations. *Proc Natl Acad Sci U S A* 108: 314-318.

[434] Joensuu H, Hohenberger P, Corless CL (2013) Gastrointestinal stromal tumour. *Lancet* 382: 973-983.

[435] Pantaleo MA, Astolfi A, Urbini M, Nannini M, Paterini P, et al. (2014) Analysis of all subunits, *SDHA*, *SDHB*, *SDHC*, *SDHD*, of the succinate dehydrogenase complex in KIT/PDGFRA wild-type GIST. *Eur J Hum Genet* 22: 32-39.

[436] Wagner AJ, Remillard SP, Zhang YX, Doyle LA, George S, et al. (2013) Loss of expression of SDHA predicts SDHA mutations in gastrointestinal stromal tumors. *Mod Pathol* 26: 289-294.

[437] Barletta JA, Hornick JL (2012) Succinate dehydrogenase-deficient tumors: diagnostic advances and clinical implications. *Adv Anat Pathol* 19: 193-203.

[438] Haller F, Moskalev EA, Faucz FR, Barthelmess S, Wiemann S, et al. (2014) Aberrant DNA hypermethylation of *SDHC*: a novel mechanism of tumor development in Carney triad. *Endocr Relat Cancer* 21: 567-577.

[439] Urbini M, Astolfi A, Indio V, Heinrich MC, Corless CL, et al. (2015) *SDHC* methylation in gastrointestinal stromal tumors (GIST): a case report. *BMC Med Genet* 16: 87.

[440] Ricketts C, Woodward ER, Killick P, Morris MR, Astuti D, et al. (2008) Germline *SDHB* mutations and familial renal cell carcinoma. *J Natl Cancer Inst* 100: 1260-1262.

[441] Paik JY, Toon CW, Benn DE, High H, Hasovitz C, et al. (2014) Renal carcinoma associated with succinate dehydrogenase B mutation: a new and unique subtype of renal carcinoma. *J Clin Oncol* 32: e10-13.

[442] Kuroda N, Yorita K, Nagasaki M, Harada Y, Ohe C, et al. (2016) Review of succinate dehydrogenase-deficient renal cell carcinoma with focus on clinical and pathobiological aspects. *Pol J Pathol* 67: 3-7.

[443] Cornejo KM, Lu M, Yang P, Wu S, Cai C, et al. (2015) Succinate dehydrogenase B: a new prognostic biomarker in clear cell renal cell carcinoma. *Hum Pathol* 46: 820-826.

[444] Gill AJ, Toon CW, Clarkson A, Sioson L, Chou A, et al. (2014) Succinate dehydrogenase deficiency is rare in pituitary adenomas. *Am J Surg Pathol* 38: 560-566.

[445] Xekouki P, Pacak K, Almeida M, Wassif CA, Rustin P, et al. (2012) Succinate dehydrogenase (SDH) D subunit (SDHD) inactivation in a growth-hormone-producing pituitary tumor: a new association for SDH? *J Clin Endocrinol Metab* 97: E357-366.

[446] Dai Z, Pan S, Chen C, Cao L, Li X, et al. (2016) Down-regulation of succinate dehydrogenase subunit B and up-regulation of pyruvate dehydrogenase kinase 1 predicts poor prognosis in recurrent nasopharyngeal carcinoma. *Tumour Biol* 37: 5145-5152.

[447] Liu J, Gao L, Zhang H, Wang D, Wang M, et al. (2013) Succinate dehydrogenase 5 (SDH5) regulates glycogen synthase kinase 3beta-beta-catenin-mediated lung cancer metastasis. *J Biol Chem* 288: 29965-29973.

[448] Chen L, Liu T, Zhang S, Zhou J, Wang Y, et al. (2014) Succinate dehydrogenase subunit B inhibits the AMPK-HIF-1α pathway in human ovarian cancer *in vitro*. *J Ovarian Res* 7: 115.

[449] Kim S, Kim DH, Jung WH, Koo JS (2013) Succinate dehydrogenase expression in breast cancer. *Springerplus* 2: 299.

[450] Ishii N, Fujii M, Hartman PS, Tsuda M, Yasuda K, et al. (1998) A mutation in succinate dehydrogenase cytochrome b causes oxidative stress and ageing in nematodes. *Nature* 394: 694-697.

[451] Huang J, Lemire BD (2009) Mutations in the *C. elegans* succinate dehydrogenase iron-sulfur subunit promote superoxide generation and premature aging. *J Mol Biol* 387: 559-569.

[452] Owens KM, Aykin-Burns N, Dayal D, Coleman MC, Domann FE, et al. (2012) Genomic instability induced by mutant succinate dehydrogenase subunit D (SDHD) is mediated by $O_2(^{-*})$ and H_2O_2. *Free Radic Biol Med* 52: 160-166.

[453] Saito Y, Ishii KA, Aita Y, Ikeda T, Kawakami Y, et al. (2016) Loss of SDHB elevates catecholamine synthesis and secretion depending on ROS production and HIF stabilization. *Neurochem Res* 41: 696-706.

[454] Safran M, Kaelin WG, Jr. (2003) HIF hydroxylation and the mammalian oxygen-sensing pathway. *J Clin Invest* 111: 779-783.

[455] Cervera AM, Apostolova N, Crespo FL, Mata M, McCreath KJ (2008) Cells silenced for SDHB expression display characteristic features of the tumor phenotype. *Cancer Res* 68: 4058-4067.

[456] Saxena N, Maio N, Crooks DR, Ricketts CJ, Yang Y, et al. (2016) SDHB-deficient cancers: the role of mutations that impair iron sulfur cluster delivery. *J Natl Cancer Inst* 108.

[457] Letouze E, Martinelli C, Loriot C, Burnichon N, Abermil N, et al. (2013) *SDH* mutations establish a hypermethylator phenotype in paraganglioma. *Cancer Cell* 23: 739-752.
[458] Mason EF, Hornick JL (2013) Succinate dehydrogenase deficiency is associated with decreased 5-hydroxymethylcytosine production in gastrointestinal stromal tumors: implications for mechanisms of tumorigenesis. *Mod Pathol* 26: 1492-1497.
[459] Xiao M, Yang H, Xu W, Ma S, Lin H, et al. (2012) Inhibition of α-KG-dependent histone and DNA demethylases by fumarate and succinate that are accumulated in mutations of *FH* and *SDH* tumor suppressors. *Genes Dev* 26: 1326-1338.
[460] Hoekstra AS, de Graaff MA, Briaire-de Bruijn IH, Ras C, Seifar RM, et al. (2015) Inactivation of *SDH* and *FH* cause loss of 5hmC and increased H3K9me3 in paraganglioma/pheochromocytoma and smooth muscle tumors. *Oncotarget* 6: 38777-38788.
[461] Gerards M, Sallevelt SC, Smeets HJ (2016) Leigh syndrome: Resolving the clinical and genetic heterogeneity paves the way for treatment options. *Mol Genet Metab* 117: 300-312.
[462] Bourgeron T, Rustin P, Chretien D, Birch-Machin M, Bourgeois M, et al. (1995) Mutation of a nuclear succinate dehydrogenase gene results in mitochondrial respiratory chain deficiency. *Nat Genet* 11: 144-149.
[463] Parfait B, Chretien D, Rotig A, Marsac C, Munnich A, et al. (2000) Compound heterozygous mutations in the flavoprotein gene of the respiratory chain complex II in a patient with Leigh syndrome. *Hum Genet* 106: 236-243.
[464] Van Coster R, Seneca S, Smet J, Van Hecke R, Gerlo E, et al. (2003) Homozygous Gly555Glu mutation in the nuclear-encoded 70 kDa flavoprotein gene causes instability of the respiratory chain complex II. *Am J Med Genet A* 120A: 13-18.
[465] Pagnamenta AT, Hargreaves IP, Duncan AJ, Taanman JW, Heales SJ, et al. (2006) Phenotypic variability of mitochondrial disease caused by a nuclear mutation in complex II. *Mol Genet Metab* 89: 214-221.
[466] Horvath R, Abicht A, Holinski-Feder E, Laner A, Gempel K, et al. (2006) Leigh syndrome caused by mutations in the flavoprotein (Fp) subunit of succinate dehydrogenase (SDHA). *J Neurol Neurosurg Psychiatry* 77: 74-76.
[467] Alston CL, Davison JE, Meloni F, van der Westhuizen FH, He L, et al. (2012) Recessive germline *SDHA* and *SDHB* mutations causing

leukodystrophy and isolated mitochondrial complex II deficiency. *J Med Genet* 49: 569-577.

[468] Ardissone A, Invernizzi F, Nasca A, Moroni I, Farina L, et al. (2015) Mitochondrial leukoencephalopathy and complex II deficiency associated with a recessive *SDHB* mutation with reduced penetrance. *Mol Genet Metab Rep* 5: 51-54.

[469] Jackson CB, Nuoffer JM, Hahn D, Prokisch H, Haberberger B, et al. (2014) Mutations in *SDHD* lead to autosomal recessive encephalomyopathy and isolated mitochondrial complex II deficiency. *J Med Genet* 51: 170-175.

[470] Ghezzi D, Goffrini P, Uziel G, Horvath R, Klopstock T, et al. (2009) *SDHAF1*, encoding a LYR complex-II specific assembly factor, is mutated in SDH-defective infantile leukoencephalopathy. *Nat Genet* 41: 654-656.

[471] Ohlenbusch A, Edvardson S, Skorpen J, Bjornstad A, Saada A, et al. (2012) Leukoencephalopathy with accumulated succinate is indicative of SDHAF1 related complex II deficiency. *Orphanet J Rare Dis* 7: 69.

[472] Van Vranken JG, Bricker DK, Dephoure N, Gygi SP, Cox JE, et al. (2014) SDHAF4 promotes mitochondrial succinate dehydrogenase activity and prevents neurodegeneration. *Cell Metab* 20: 241-252.

[473] Parvari R, Levitas A (2012) The mutations associated with dilated cardiomyopathy. *Biochem Res Int* 2012: 639250.

[474] Davili Z, Johar S, Hughes C, Kveselis D, Hoo J (2007) Succinate dehydrogenase deficiency associated with dilated cardiomyopathy and ventricular noncompaction. *Eur J Pediatr* 166: 867-870.

[475] Levitas A, Muhammad E, Harel G, Saada A, Caspi VC, et al. (2010) Familial neonatal isolated cardiomyopathy caused by a mutation in the flavoprotein subunit of succinate dehydrogenase. *Eur J Hum Genet* 18: 1160-1165.

[476] Alston CL, Ceccatelli Berti C, Blakely EL, Olahova M, He L, et al. (2015) A recessive homozygous p.Asp92Gly SDHD mutation causes prenatal cardiomyopathy and a severe mitochondrial complex II deficiency. *Hum Genet* 134: 869-879.

[477] Lu M, Zhou L, Stanley WC, Cabrera ME, Saidel GM, et al. (2008) Role of the malate-aspartate shuttle on the metabolic response to myocardial ischemia. *J Theor Biol* 254: 466-475.

[478] Wang Y, Li C, Chuo W, Liu Z, Ouyang Y, et al. (2013) Integrated proteomic and metabolomic analysis reveals the NADH-mediated TCA

cycle and energy metabolism disorders based on a new model of chronic progressive heart failure. *Mol Biosyst* 9: 3135-3145.

[479] Wu JL, Wu QP, Huang JM, Chen R, Cai M, et al. (2007) Effects of L-malate on physical stamina and activities of enzymes related to the malate-aspartate shuttle in liver of mice. *Physiol Res* 56: 213-220.

[480] Cascon A, Comino-Mendez I, Curras-Freixes M, de Cubas AA, Contreras L, et al. (2015) Whole-exome sequencing identifies MDH2 as a new familial paraganglioma gene. *J Natl Cancer Inst* 107.

[481] Lo YW, Lin ST, Chang SJ, Chan CH, Lyu KW, et al. (2015) Mitochondrial proteomics with siRNA knockdown to reveal ACAT1 and MDH2 in the development of doxorubicin-resistant uterine cancer. *J Cell Mol Med* 19: 744-759.

[482] Liu Q, Harvey CT, Geng H, Xue C, Chen V, et al. (2013) Malate dehydrogenase 2 confers docetaxel resistance via regulations of JNK signaling and oxidative metabolism. *Prostate* 73: 1028-1037.

[483] Liu Y, Asnani A, Zou L, Bentley VL, Yu M, et al. (2014) Visnagin protects against doxorubicin-induced cardiomyopathy through modulation of mitochondrial malate dehydrogenase. *Sci Transl Med* 6: 266ra170.

[484] Mali Y, Zisapel N (2009) A novel decoy that interrupts G93A-superoxide dismutase gain of interaction with malate dehydrogenase improves survival in an amyotrophic lateral sclerosis cell model. *J Med Chem* 52: 5442-5448.

[485] Shi Q, Gibson GE (2011) Up-regulation of the mitochondrial malate dehydrogenase by oxidative stress is mediated by miR-743a. *J Neurochem* 118: 440-448.

[486] Oh SJ, Gu DR, Jin SH, Park KH, Lee SH (2016) Cytosolic malate dehydrogenase regulates RANKL-mediated osteoclastogenesis via AMPK/c-Fos/NFATc1 signaling. *Biochem Biophys Res Commun* 475: 125-132.

[487] Lee SM, Kim JH, Cho EJ, Youn HD (2009) A nucleocytoplasmic malate dehydrogenase regulates p53 transcriptional activity in response to metabolic stress. *Cell Death Differ* 16: 738-748.

[488] Lee SM, Dho SH, Ju SK, Maeng JS, Kim JY, et al. (2012) Cytosolic malate dehydrogenase regulates senescence in human fibroblasts. *Biogerontology* 13: 525-536.

INDEX

#

A

B

C

D

E

G

L

M

N

O

P

R

S

T

V

W

X

α